THE WORLD OF SCIENCE

COLD-BLOODED ANIMALS

THE WORLD OF SCIENCE

COLD-BLOODED ANIMALS

MAURICE BURTON

Facts On File Publications
New York, New York • Bicester, England

COLD-BLOODED ANIMALS

First published in the United States of America in 1986
by Facts on File, Inc., 460 Park Avenue South, New
York, N.Y. 10016

First published in Great Britain in 1986 by
Orbis Book Publishing Corporation Limited, London
A member of Maxwell Pergamon Publishing Corporation plc

Library of Congress Cataloging in Publication Data
Main entry under title:
World of Science

Includes index.
Summary: A twenty-five volume encyclopedia of
scientific subjects, designed for eight- to twelve-year-
olds.
One volume is entirely devoted to projects.
1. Science—Dictionaries, Juvenile. 1. Science—
Dictionaries
Q121.J86 1984 500 84-1654

ISBN: 0-8160-1074-9

Printed in Italy
10 9 8 7 6 5 4

Consultant editors
Eleanor Felder, Former Managing Editor, **New Book of Knowledge**
James Neujahr, Dean of the School of Education, City College of New York
Ethan Signer, Professor of Biology, Massachusetts Institute of Technology
J. Tuzo Wilson, Director General, Ontario Science Centre

Previous pages
Garden spider
carrying her prey.

Editor Penny Clarke
Designer Roger Kohn

CONTENTS

1 WHAT IS A COLD-BLOODED ANIMAL?

2 SIMPLE INVERTEBRATES

3 MOLLUSCS AND ECHINODERMS

4 ARTHROPODS – ANIMALS WITH JOINTED LEGS

5 SHARKS, RAYS AND BONY FISH

6 AMPHIBIANS

7 REPTILES

8 WARM-BLOODED OR COLD-BLOODED?

Note There are some unusual words in this book. They are explained in the Glossary on pages 62–63. The first time each word is used in the text it is printed in *italics*.

▼ In some kinds of coral, parts of the colony, such as these, catch food, while other parts digest it.

1 WHAT IS A COLD-BLOODED ANIMAL?

A COLD-BLOODED ANIMAL IS...

All animals are affected by the temperature of their surroundings, the places in which they have to live. If these get too hot or too cold, the animal may become ill or even die. But many animals have learned to live in hot or cold places, from inhospitable deserts and steamy jungles to icy wastes such as those of Siberia and the polar regions.

If you think of the kinds of animals that can endure, for example, extreme cold, you will come up with such names as polar bear, penguin, seal, and perhaps snow goose, blue whale or even icefish. As it happens, all but one of these animals is either a mammal or a bird, that is, a warm-blooded animal. Thanks to the protection of their fur, feathers or blubber, such warm-blooded creatures are able to endure extreme cold and remain active. On the other hand, a lizard or other reptile, with none of these protections, would soon freeze to death in polar weather. Its body and blood temperature would drop to the temperature of its surroundings, too low for its life to continue.

But what of the icefish? This is both a

▼ The blue-tongued skink of Australia is a reptile and, like many other cold-blooded animals that live in places with a hot climate, it spends the hottest part of the day out of the sun, otherwise it might die of heat-stroke.

◀ One of the many types of anglerfish that live in some of the deepest parts of the ocean, where the water temperature is very low.

▲ Siphonophores are relatives of the jellyfish. They drift at the surface of the sea, moved around by tides and currents.

polar and a cold-blooded animal. Its body is always at the same, near-freezing temperature as the seawater in which it lives. Like other cold-blooded, icy-water animals, the icefish is a very sluggish creature. Most mammals and birds, on the other hand, are much more active creatures, and so are many of the cold-blooded animals living in warmer climates. In these places the problem can be too much body heat rather than too little. A mammal, overheated by sun or hard running, will get rid of its extra body heat by sweating or panting. A lizard or other cold-blooded animal, unable to stop its body temperature rising in the hot sun, will seek a shady place or burrow its way out of the sunlight, to remain cool. In the sharp cold of a desert night, a lizard, like the icefish, becomes a very sluggish creature.

Exceptions to test a rule

To some extent, the words cold-blooded and warm-blooded are misleading. For instance, some warm-blooded animals find the cold of winter or the cold nights in a hot desert, dangerous. So, to avoid the problem, they go to sleep. At the same time they let their heat-control go out of action. A warm-blooded animal, such as a woodchuck or dormouse, that does this in winter is said to hibernate. Some people call this *hibernation* 'winter sleep'.

Cold-blooded animals also slow down in winter. They are said to become *torpid*. They also become torpid if the weather suddenly turns cold in summer.

Yet other animals can be thought of as cold-blooded even though they have no blood! These are the many kinds of small or tiny animals that live in water, some consisting of only a single living cell, others of many cells, but all far simpler than the animals we have mentioned so far. The next pages deal with these 'cold-blooded' creatures, before we come to earthworms – which really do have blood!

▲ A two-tailed pasha butterfly clings to the remains of the pupa from which it has just emerged. It will wait until its wings expand to their full size before flying off. Like all other insects it is cold-blooded. There is a picture of the caterpillar of this butterfly on page 34.

PROTOZOA – SINGLE-CELLED ANIMALS

The bodies of most animals are made up of large numbers of cells. There are, however, thousands of species that are known as Protozoa, or first animals, from the Greek words *protos* meaning 'first' and *zoon* meaning 'animal'. They are called first animals because it is believed that they were the first animals to come into being. Another name for Protozoa is single-celled animals.

All protozoa are small, but most are microscopic and cannot be seen with the naked eye. They live in moist or wet places, in the sea or in fresh water, or in the bodies of larger animals.

Although there are many different kinds of protozoa they are all alike in that every function of the body goes on inside a single cell. The body has no organs such as brain, nerves, heart, lungs or stomach. All the duties performed by these organs in the body of a multi-cellular animal, such as a dog, a cat, a lion or a human being, are carried on in this tiny blob of living protoplasm. There are four main groups of protozoa: amoebae, flagellates, sporozoa and ciliates.

▼ *Volvox* colonies at the surface of a pond. Each colony usually consists of hundreds of individual organisms. A large colony is about the size of one of the full stops in this book.

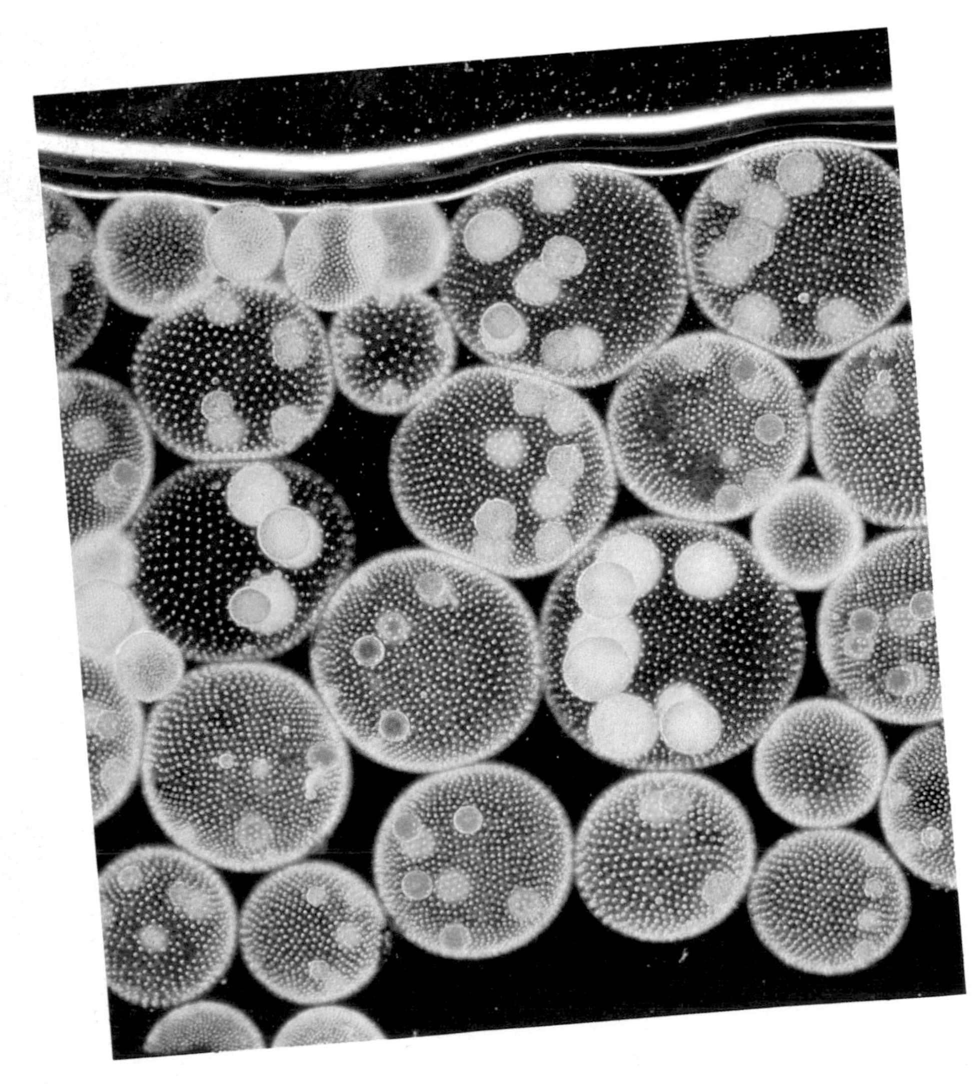

Amoeba

The best known of these groups is amoeba. Its body consists of a single cell – a small piece of *protoplasm* with a central nucleus. The nucleus controls everything that goes on inside the cell. When an amoeba moves, parts of its body flow out into finger-shaped pseudopodia or false feet.

The body of an amoeba is naked. When moving it changes shape and seems to flow along. There are relatives of amoeba that are not naked. They build themselves a case of very tiny sand grains or make themselves a shell of chalk or of a glass-like material known as silica. Their pseudopodia are like long threads and protrude through tiny holes all over the shells.

Flagellates

The word flagellate means 'whip-bearer', and the protozoa known as flagellates have one or more long, very thin hairs, known as *flagella*, on their bodies. These hairs thresh the water like the lash of a whip and propel the animal through the water.

Sporozoa

The third class, the sporozoa, are all *parasites* and live in the bodies of larger animals. Some cause diseases, such as malaria.

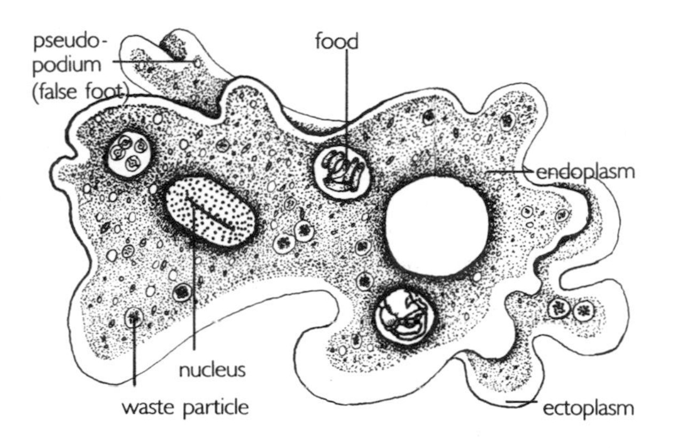

◀ A drawing of what the scientist sees in amoeba under the microscope, as a result of long and careful study. The nucleus controls all the animal's activities. Inside the tough ectoplasm, or skin, is the soft endoplasm. This is the real, living part of this single-celled animal.

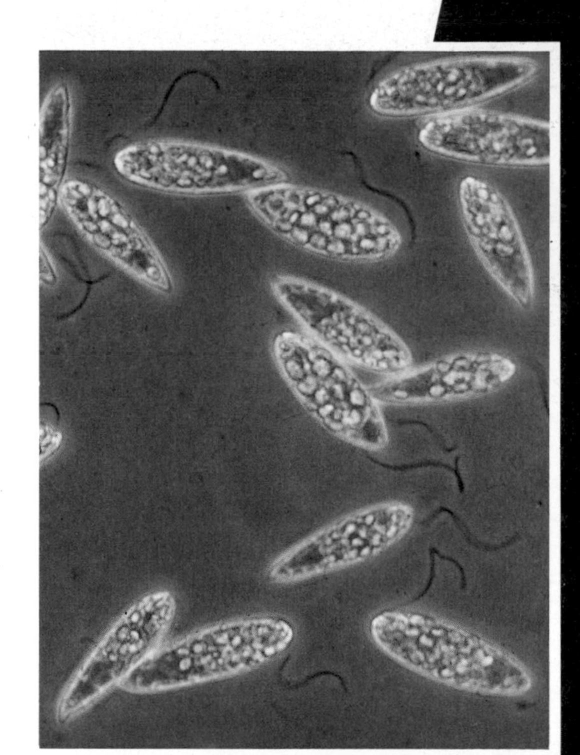

► You can just see the long whip-like thread, or flagellum, on each of these tiny flagellates called *Euglena*.

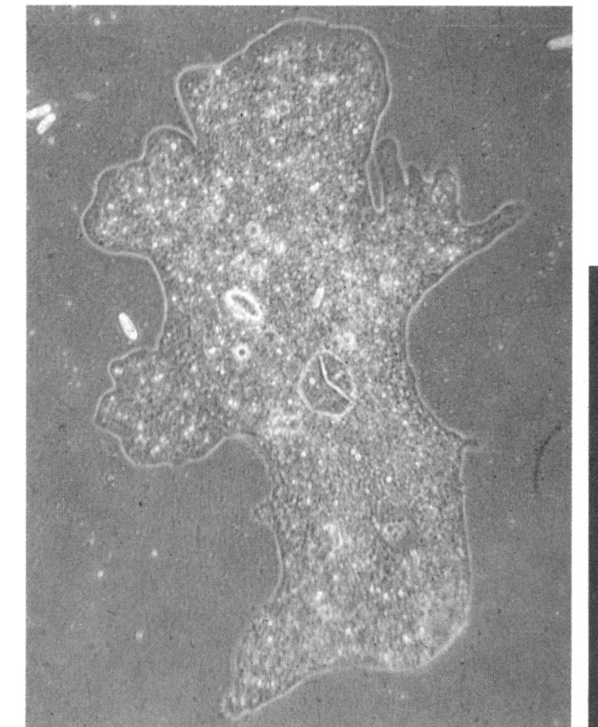

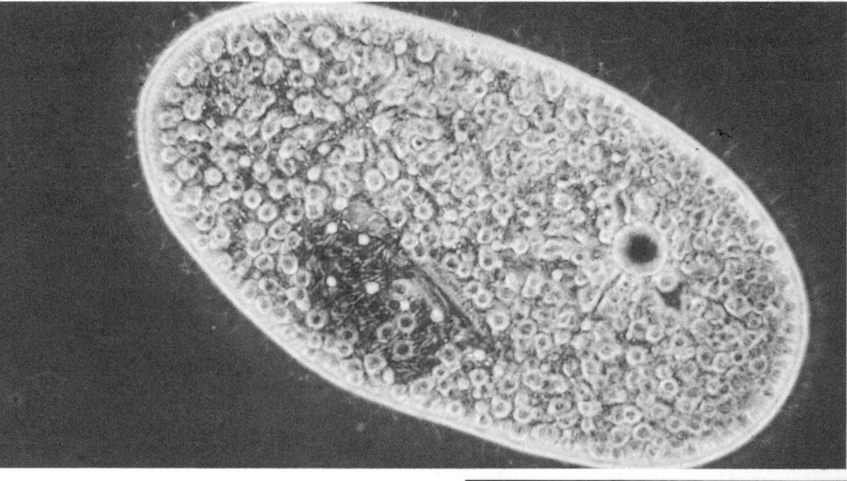

▲ Amoeba, as seen under the microscope. The whole animal consists of a single cell in which all the processes of life go on. To all appearances it is no more than a tiny blob of clear jelly.

► This radiolarian is also a member of the group of single-celled animals called Protozoa. Unlike most of the others it has a stiff outer 'skeleton'.

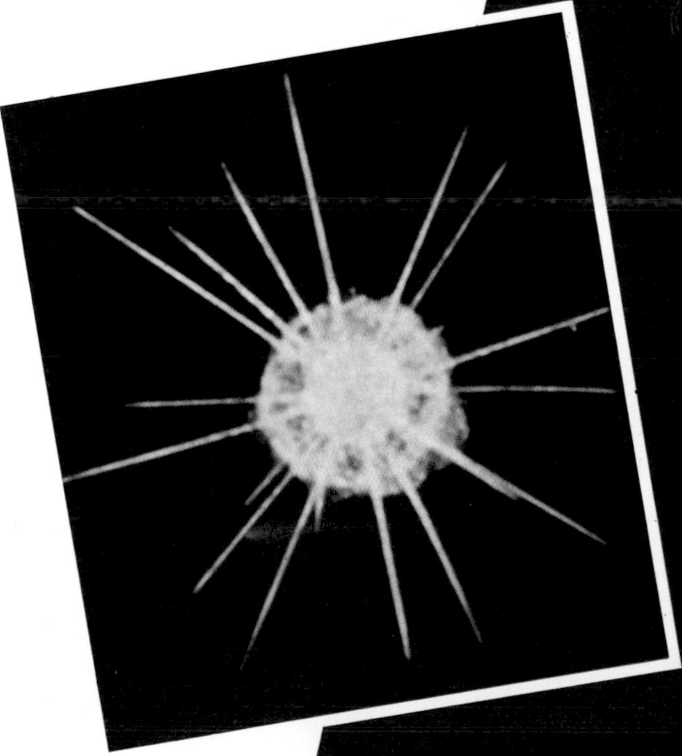

▲ Paramecium, a single-celled animal which, unlike amoeba, keeps its shape. It swims through water, driven by the beating of the tiny cilia on its surface. These act like the numerous oars in a slave galley.

Ciliates

The last class of Protozoa, the ciliates or hairbearers, get their name from the many short hairs projecting from their bodies. These are not true hairs, like those on our heads, but like flagella, are much thinner. The *cilia* move in waves like the tops of long grass in a field when a gentle breeze blows over it. The wave-like movements of the cilia makes the body move through water.

SPONGES

There are around 3,000 different types of sponge. The best-known are the half-dozen species we call bath sponges. The rest, although of no economic value, are immensely important in helping to keep clean the seas, lakes and rivers everywhere in the world, because they feed on bacteria and tiny particles of dead plants and animals.

In contrast with the protozoa, sponges are much more complex organisms. Each sponge is made up of many different cells, and there are different kinds of cell to carry out different functions. No longer does one cell do everything.

Sponges may look like plants, but they are animals that are fixed in one place. Some are branching and green like plants. Others are cake-shaped or fan-shaped. Many are brightly coloured in reds and purples, others are a dull brown. Whatever their shape or colour, all sponges have their surface covered with tiny holes known as pores. They also have groups of larger holes, like craters. Water is drawn in through the pores, pumped through the body and driven out through the craters, rather like a tiny pumping station. The incoming water brings food and oxygen and the outgoing water carries away waste matter.

The inside of the sponge's body is a labyrinth of tiny tubes which balloon out at some point along their length. The inside of each balloon is lined with special cells. Each cell has a whip-like *flagellum*,

▼ Three-rayed spicules of carbonate of lime that make up the skeleton of a purse sponge.

► A sponge growing on a coral reef in the warm seas around the Bahamas. The sponges that grow in shallow tropical seas are usually brightly coloured.

just like that of a protozoa with flagella (page 8). But in the sponge, the lashing flagella do not propel the animal along, instead they cause small currents of water to flow from the pores to the crater. It is these flagellated cells that make a sponge work like a pumping station.

Like plants, sponges have no sense-organs, neither do they have a heart, blood-vessels, brain or stomach. Some of them have a few muscle-fibres and simple nerves. Unlike plants they reproduce by means of swimming larvae; and they have no cellulose, the woody material peculiar to plants.

Different types of sponge

Most sponges, such as the purse sponge, have skeletons made up of tiny spicules (little spikes). The rest have skeletons made up of a network of thin fibres, similar to silk but tougher. The bath sponge is one of these.

Sponges of different kinds are found everywhere in the world, in lakes, rivers and the sea, from the equator to the poles. In the sea they are found from near high-tide mark to the greatest depths of the oceans. Wherever they live they are firmly anchored to the surface of rocks or seaweeds.

▲ The two different types of hole found in sponges are clearly visible on this branched sponge. On the sides of each finger of sponge are the small holes or pores through which water enters the sponge. At the top of each branch are the large crater-like holes through which water leaves the colony.

NETTLE ANIMALS

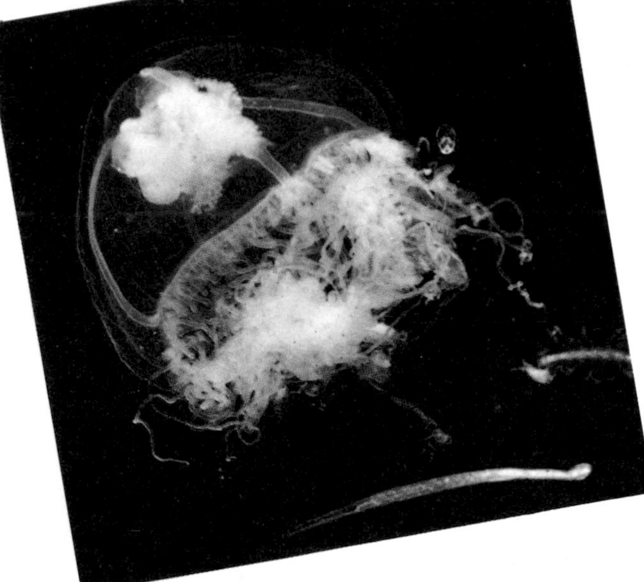

▲ Three tiny sea-anemones that have budded from the adult beside them. The budding young remain attached to the stomach of the parent (see the hydra on page 13) until they are fully developed. Then they detach themselves and lead quite separate lives.

▲ This jellyfish has caught two arrow-worms. It has paralyzed them with its stings. Later it will contract its slender tentacles to draw the arrow-worms up to its mouth.

You would never think, merely from looking at them, that a sea-anemone and a jellyfish are close relatives. Yet both belong to a class known as nettle animals, because they sting. Usually if you are stung by one of these creatures all that happens is that you get a rash, but a few sting so badly that people have died as a result.

Most nettle animals live in the sea. A few live in fresh water, in lakes or rivers. Corals are nettle animals, so are sea-firs and sea-fans. Some of the sea-firs are so small they look more like tiny mosses. Others, like the brain coral, may be 1 m (3 ft) or more across and nearly the same high. The jellyfishes may be no bigger than a pin's head or they may be nearly 2 m (6 ft) across.

A nettle animal is either jelly-fish shaped or shaped like a *polyp*. This Greek word was first used to describe an octopus, which has many arms, or tentacles, surrounding its mouth. A sea-anemone also has many tentacles surrounding its mouth. The first scientists noted this similarity and called the sea-anemone a polyp. Since then the word 'polyp' has been used especially for nettle animals.

Effective weapons

A hungry sea-anemone polyp waves its tentacles about until one of them touches a fish, shrimp or other small animal. When this happens, some of the thousands of stinging cells in the skin of the tentacle are shot out. A stinging cell is made up of a bag of poison and a long hollow thread. When a thread pierces the skin of another animal it holds that animal, if it is small. Then poison flows down the hollow thread. If the animal is small it is paralyzed and dies. Then it is eaten. If it is large, such as a human swimmer brushing against a large sea-anemone or a jellyfish, it may suffer a painful rash.

Apart from their stinging cells nettle animals are very simple. They have no brain or blood, very few nerves, simple muscles, no heart, liver or other digestive organs. Each polyp is little more than a bag with a mouth at one end surrounded by tentacles. In anemones the mouth is at the upper end. In a jellyfish it is on the underside. When food is caught the tentacles carry it to the mouth. The nettle animal then rests and digests it. What cannot be digested is then passed out of the body through the mouth.

Jellyfishes spend their lives swimming. Polyps, such as the anemones, spend their time stuck to rocks or other solid objects. Some polyps can build a solid skeleton. This may be hard and chalky, as in corals, or it may be horny, as in sea-fans. So corals, sea-firs and sea-fans are really colonies of many polyps supported by their solid skeletons.

► A few nettle animals live in lakes and rivers. One is called hydra. The hydra shown here is about 2.5 cm (1 in) long. It is budding off a young hydra. Hydra consists of a single polyp.

◄ This nettle animal is called a sea-fan. It lives in the warmer waters of the world. Like the red coral it is a colony of hundreds of small polyps which live on its branches.

▲ Two different types of coral and several small sponges growing on the Great Barrier Reef off the east coast of Australia.

► Close-up of one of the polyps of the red coral that lives in the Mediterranean Sea. Its branches are colonies of polyps. Each polyp has eight tentacles and looks like a very small sea-anemone.

EARTHWORMS

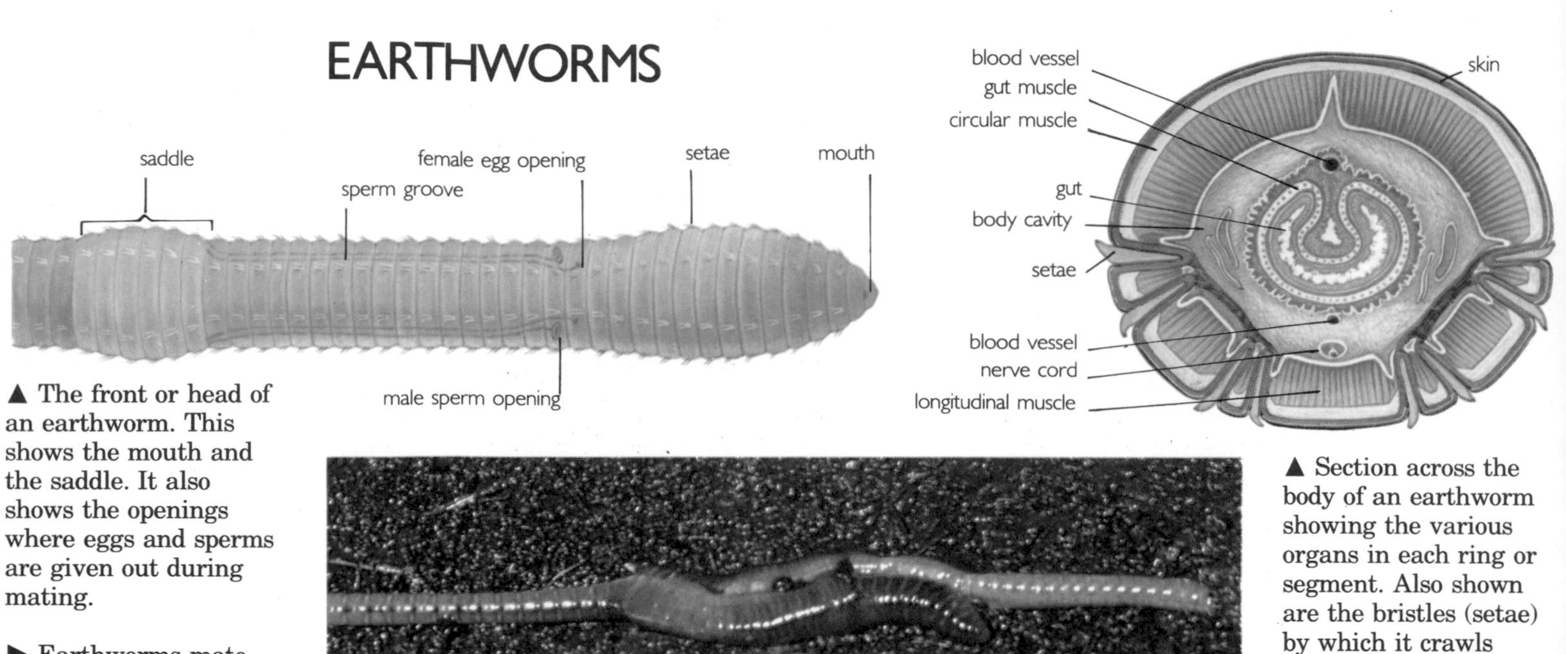

▲ The front or head of an earthworm. This shows the mouth and the saddle. It also shows the openings where eggs and sperms are given out during mating.

► Earthworms mate at night, reaching out to each other from their burrows. An earthworm is an hermaphrodite. That is, it is both male and female in one. But an earthworm is unable to fertilize itself. While the two worms are lying side-by-side they exchange sperms so the eggs in each worm are fertilized.

▲ Section across the body of an earthworm showing the various organs in each ring or segment. Also shown are the bristles (setae) by which it crawls about.

You may have seen earthworms crawling over the ground after rain, or found them if you have dug in the ground. Usually they are no longer than an ordinary pencil, and may be much less. They hide under stones where it is damp or among dead leaves, but mainly they burrow in the ground. Direct sunlight soon kills earthworms but at night, especially if there has been rain, they come out onto the surface and crawl around. It is then, if you keep very still and silent, that you can learn how they are able to crawl. If your hearing is good and you listen carefully, you may hear a very faint scratching sound. This is made by the bristles in their skin as they use them to crawl. These bristles, or *setae*, grow in pairs on the lower part of each ring or segment of the body.

Earthworms not only live in earth, they eat it. They swallow it and as the earth passes through the long tubular gut, any small particles of decaying plants and animals in the earth are digested. The earth then passes out of the body as coiled 'strings', known as worm casts. These are easy to spot on lawns in gardens or parks.

The earthworm's blood system consists of fine blood-vessels and includes ten hearts! A nerve cord lying under the gut runs from end to end of the body. At the head end the cord is enlarged and forms a sort of brain.

About a third of the way from the head end the body is slightly swollen. This part is known as the saddle. You may have been told that this is where the worm has been cut in two and grown fresh parts. Worms can do this, but the saddle is quite different. It is made up of glands for secreting a tubular cocoon in which eggs are laid. The worm then withdraws its body from the cocoon, leaving the eggs to hatch into baby earthworms.

How long is long?
Although the majority of earthworms may be only a few centimetres long, there are giants in hot countries. In Australia, for instance, these may be 4 m (12 ft) long when fully stretched. Earthworms are rather similar to a piece of elastic. This is because they have two sets of muscles. One set, called longitudinal muscles, run the whole length of the body. Another set, called transverse muscles, run round the body. When the transverse muscles contract the body lengthens. When the longitudinal muscles contract the body shortens. It is these wave-like movements of lengthening and shortening, together with the grip of the setae or bristles, that cause the earthworm to move along. Because an earthworm has no skeleton it is these muscles which give its body strength as well as elasticity. In fact an earthworm can tie itself in knots – you sometimes find them like this in the earth or under a large stone.

▼ Sponges and nettle animals have no blood or blood-vessels. The earthworm is the lowliest animal to have these. There is a main blood vessel running the length of the body above the intestine and another beneath it. There are five pairs of small hearts connecting the worm's ten hearts.

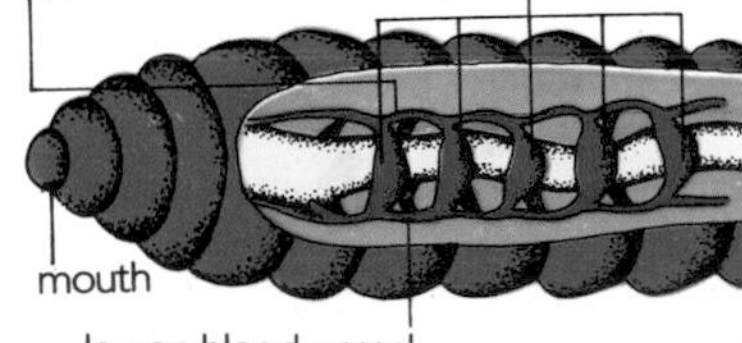

MARINE BRISTLE WORMS

Worms do not live only in the ground. There are many different kinds that live in the world's seas and oceans, from the tropics to the coldest seas, from high up on the shore to the greatest depths of the oceans. There are a few species that live in salt marshes.

These marine bristle worms, as they are called, are of two kinds. There are those, such as the ragworms, that move about from place to place. Then there are those such as the fanworms and the peacock worms, that stay in one place once they have changed from the *larva* to the adult.

A wandering life

Marine worms that move around have small bristles all along the sides of their bodies, just as earthworms do. But ragworms have several bristles to each segment or ring. These bristles stick out from the ends of parapodia, or short, flap-like legs, two to each ring of the body. The wandering marine worms crawl through mud or sand or into crevices in rocks, using the bristles on their legs to get a grip on the surface over which they are crawling and push themselves along.

Their food, which they eat with strong chewing jaws, consists of small dead sea creatures. So they are useful scavengers on the sea bed.

▲ This fanworm is sometimes called the feather-duster worm. The feather-like head is used for breathing and for catching small particles of food floating in the water. It can be quickly withdrawn for safety into the tube of sandgrains the worm builds for itself.

▼ Marine bristleworms like this one are sometimes called ragworms. They have tufts of bristles on each side of the body, which give it a ragged look. Ragworms have sharp jaws and eat other small animals.

The tube-dwellers

Fanworms and peacock worms look quite different to the ragworms. They remain fixed in one place for life and make a tube in which to shelter. This tube may be made of grains of mud or sand or it may be made of a substance given out from the worm's own body, which is chalky, horn-like or jelly-like. These tubes can be seen on rocks or shells, they are small and hard and sometimes coiled.

From its tubular shelter the worm puts out a crown of feathery, beautifully patterned tentacles. The worm uses its tentacles to catch small particles of food, and also, if its tube is of mud or sand, to build the tube itself.

A tube-dwelling worm withdraws into its tube to rest or if it is disturbed. When hungry it climbs up the inside of the tube and spreads out the crown of tentacles which surrounds the mouth. Then it becomes obvious why the fanworm got its name, because the crown of tentacles opens like a fan. It also closes like a fan as the worm withdraws into its tube when it has ceased feeding or is disturbed.

Another species of marine worm lives in crevices in rock. It too, has a crown of tentacles, but they are not arranged in a fan. The tentacles are numerous, red, and very long and slender. When it is feeding these long thin tentacles can be seen spreading out well beyond the hole in the rock in which the worm is living, but as soon as it is disturbed the threads are withdrawn into the hole in which the worm is living.

▲This type of marine bristleworm grows up to 25 cm (9¾ in) in length, with between 90 and 100 segments to its body. It makes twisted burrows in the sand or mud near the low water mark on a beach.

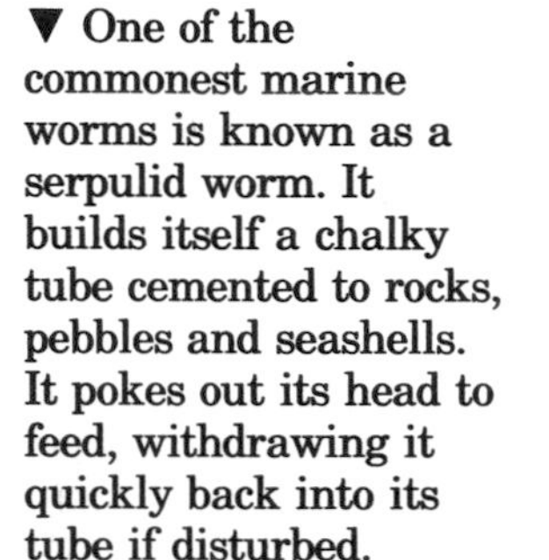

▼ One of the commonest marine worms is known as a serpulid worm. It builds itself a chalky tube cemented to rocks, pebbles and seashells. It pokes out its head to feed, withdrawing it quickly back into its tube if disturbed.

► Lugworms' holes have a depression at one end where the worm has taken in sand or mud. At the other is a cast of the sand the worm has expelled.

LEECHES

Leeches are related to earthworms. They have the same type of ringed body but they have no bristles. Some live in water and swim by undulating their bodies. Others live on land and crawl by looping along in a similar way to many caterpillars. A leech has two suckers by which it can hold onto a solid support. One is at the front of the body, the other is at the hind end. The leech 'loops' by gripping on with the back sucker, stretching the body forward, gripping with the front sucker, then releasing the back one and bringing it up to the front one and taking a grip. Then it releases the front sucker and the process continues.

The best-known species is the medicinal leech. This was used by doctors years ago in blood-letting. Blood-letting was thought to give relief from pain and, usually wrongly, to help in the cure of almost any illness. A leech was placed on a patient's skin where it held fast with its suckers. Then it pierced the skin with the three sharp teeth in its mouth and started to suck the blood that oozed from the wound.

A leech can take as much as five times its own weight of blood at a meal. It may take as much as 20 minutes to gorge itself. After that, it can, if necessary fast for a year. When hungry a leech is cylindrical, but when it is fully fed it becomes fat and bloated.

Leeches too are found throughout the world. Some live in the sea, others in rivers and lakes or among damp vegetation. They are most common in the tropics. Those that live on land are found on bushes and trees. It is almost impossible to put a foot to the ground in jungles and marshes without being attacked by leeches. They creep up the legs or drop from trees and bushes onto the head and neck of a passer-by.

Hungry parasites

The leech knows when a person or a warm-blooded animal goes by, because it can detect the heat of its body. It reaches out to get a hold with its sucker, even springing a short distance to reach its victim. When leeches living in water sense a warm-blooded animal going by, they swim up and fix themselves to it.

Not all leeches feed as parasites on the bodies of much larger, warm-blooded animals. Some feed on earthworms, insect larvae and snails, attacking them and sucking the fluids from their bodies.

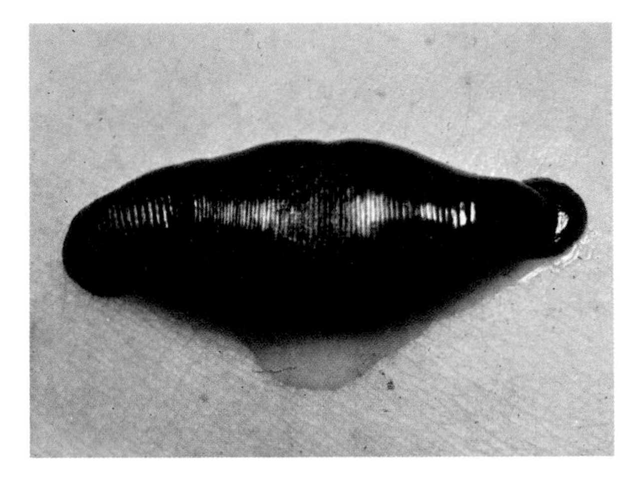

▲ 'To cling like a leech' is a common saying. This leech is clinging to human skin by its rear sucker (left), while holding on by its front sucker to suck blood.

▼ A medicinal leech of the kind commonly used by doctors to bleed their patients. Blood-letting of this kind is rarely used today. The leech shown here is 12.5 cm (5 in) long when fully extended.

3 MOLLUSCS AND ECHINODERMS

BIVALVE MOLLUSCS

▼ A scallop with its shell wide open for breathing and feeding. By snapping shut its shell it can drive itself rapidly through the water. The small gleaming dots are its eyes. **Inset** Close-up of an eye and the tentacles at the edges of the mantle.

It was Aristotle, the scholar and zoologist who lived in Greece from 384 to 322 BC, who first used the word *mollusc*. He used the name to describe many different kinds of soft-bodied animal. Other scholars accepted his ideas until the nineteenth century. Then the use of the name was restricted to several classes of animals the chief of which are bivalves, slugs and snails, octopuses and cuttlefishes.

Bivalves are shellfish living mostly in the sea, but also in lakes and rivers. They have a soft body enclosed in a shell and the shell is in two parts, or valves, that are hinged. ('Bi' comes from a Latin word meaning two.)

Today the word 'valve' is used for many different things quite unrelated to shellfish, so its use may seem odd at first. But if we go back to Roman times, we find that the Romans used it for each of the two halves of a folding door. And so the use of the word makes sense – especially as the Romans regarded many shellfish as good to eat.

Probably the best known today of the bivalves is the oyster. It is famous both as the producer of pearls and as a table delicacy. There are other bivalves, also eaten as delicacies, such as mussels, clams, scallops and cockles. Although they differ in shape, size and colour and in the places where they live, all these consist of a soft body enclosed in a shell made up of two valves or parts.

Different but the same

From giant clams to tiny cockles, all bivalves have the same basic anatomy or body plan. The soft body is made up of a number of flaps, like the pages of a book. On the outside of the body, almost like the covers of a book, is a fold of flesh known as the *mantle*. From the mantle comes the chalk that forms the animal's shell.

Inside the mantle are flat folds of flesh with many tiny holes or slits in them. They are the *gills*. And between the gills is the flat, fleshy body containing the stomach, intestine and other organs.

Bivalves feed by drawing water into the shell and pumping it out again. As the water passes through they filter out any particles of food. Microscopic plants and animals or small particles of dead plants and animals are sieved from the water and swallowed. At the same time oxygen is extracted from the water.

The largest bivalves live in the tropical waters of the Indian Ocean. They may be 1 m (3 ft) across and weigh 130 kg (300 lb). They, like many other bivalves, simply lie on the bottom of the sea or river, usually buried or half-buried in the sand or mud. A few, such as the cockle, have a muscular foot with which they can jump to escape enemies, the chief of which are starfishes. Others merely bury themselves deeper in the sand to escape being eaten. The scallops can, by opening and closing the two parts of their shell, swim short distances.

▲ A giant clam on a coral reef in the seas of Indonesia. On the left can be seen an opening in the mantle. This is a siphon through which water, carrying oxygen and food, is drawn in.

▼ A cockle lying on its side. The two hairy siphons show up white. The one on the left is drawing in water containing oxygen and food. From the right-hand siphon a string of waste matter is being carried out.

▲ A spiny cockle: its two siphons can be seen. Water sucked in through one siphon brings food and oxygen. Waste matter is driven out through the other siphon.

SLUGS AND SNAILS

► The familiar shape of a land slug. It crawls on a fleshy undersurface known as its foot. Glands in the foot give out a slime which enables the slug to move easily over almost any surface.

◄ A batch of slug's eggs somewhat magnified. They look like pearls and are laid in damp, dark places, usually in soft soil.

These two kinds of animal are closely related. The main difference between them is that a snail carries with it a shell of its own making in which it can shelter. Both are known as gastropod molluscs. The word 'gastropod' means 'stomach-footed'.

The foot, in both slugs and snails, is the fleshy part that rests on the ground and on which they crawl about. In front is the head with a mouth. Above the mouth are two pairs of tentacles, one pair above the other. The upper pair end in eyes, the lower, smaller, pair have other sense organs. Both pairs of tentacles can be withdrawn when not in use.

We are all familiar with the slugs and snails that live on land. They crawl about on their fleshy foot and eat dead plants. Sometimes they eat living plants. This is why they are looked upon as enemies of the gardener. In fact, they do a great deal of good because they help compost dead and rotting vegetation, turning it into humus that enriches the soil.

On the upper surface of a slug, on the front half of the body, is a sort of hump. This is called the mantle. Snails also have mantles, but in a snail it is hidden by the shell. One of the uses of the mantle is that it produces the lime from which the shell is made. Some kinds of slug also have a shell, although it is only a very small one. Others have a shell hidden under the skin.

The mantle is also used for breathing. Inside it is a cavity that acts as a lung. The inside wall of the mantle is, like the wall of a lung, very rich in blood-vessels. Air is taken in through a small breathing hole at the side of the mantle, where it joins the fleshy part of the foot.

A slug or snail has a kind of tongue, known as a *radula*. This is a horny ribbon, its surface is coated with hundreds of horny teeth. It acts in the same way as a file as the animal works it backwards and forwards to rasp away pieces of food which are then swallowed. Although most snails and slugs eat plant food, some slugs

► A banded snail of Europe, common in gardens and on farms. It has been accidentally introduced into North America.

▼ A land snail laying a batch of eggs. One egg can be seen leaving the body near the snail's right-hand stalked eye.

►One of the many species of beautiful sea-slug. The red finger-like growths on the body are the gills through which the sea-slug breathes.

▼ This species of sea-slug has its gills in a tuft at the rear end of the body.

are carnivorous. They feed on other slugs or on earthworms.

Land slugs and land snails are very numerous throughout the world. There are even snails that live in deserts – they survive the long dry periods by burrowing down to where the earth is moist. There are also snails that live in ponds and rivers. But the most numerous, and certainly the largest, are the sea-slugs and sea-snails. They are also much more brightly coloured. They feed on seaweeds, sponges and nettle animals. Among the many sea-snails are periwinkles, usually called winkles, whelks, limpets and earshells. Sea-snails and sea-slugs usually breathe by gills.

◄ The sea hare is yet another type of sea-slug. It lives in the Mediterranean Sea and is here feeding on its main food – sea lettuce.

OCTOPUS AND SQUID

The soft-bodied molluscs come in many shapes and sizes. It is hard to believe that the octopus, that fabled monster of the deep, is second cousin to slugs and snails. The pictures you sometimes see of a diver fighting for his life coiled in the writhing tentacles of a giant octopus are far from anything a snail or a slug could achieve, so they seem as if they are entirely different kinds of animals.

It is even harder to see a close link between slugs and snails and squids. The first two move slowly, 'at a snail's pace', as the saying goes. A squid, by contrast, moves swiftly, sometimes even leaping out of the water and landing on the deck of a large ship.

It may be that stories of tall ships dragged to the bottom of the sea by giant squid are far-fetched. Yet the size of the largest-known giant squid makes them just possible. Nobody knows exactly how long the biggest squid is, but a specimen of nearly 30 m (100 ft) including its two long tentacles, has been measured.

The octopus has a rounded body with a mouth in front surrounded by eight arms bearing rows of suckers. The mouth is furnished with a pair of powerful horny jaws, similar to a parrot's beak. These are used to crunch up the bodies of crabs and other prey.

Underneath the body is a tube, known as a siphon. This leads into a mantle cavity containing gills. Sea water is drawn into the mantle cavity and squirted out again. That is how the animal breathes. When the octopus needs to travel, to pounce on prey or escape enemies, water is taken in and driven out with such force that it drives the octopus backwards.

Octopuses and squids also have an ink-sac which opens into the intestine. When threatened, both shoot a dark 'ink' out through the siphon. This baffles their enemies and allows the threatened animals time to escape.

Inner and outer shells

The octopus does have a kind of shell but it is inside the body. In cuttlefishes, near relatives of the octopus, what remains of the shell is known as cuttlebone. In squids the shell is just a horny 'pen', shaped like an old-fashioned fountain pen-nib. Other relatives of the octopus and squid are the pearly nautilus and the paper nautilus. They have coiled shells that look more like the shells of other molluscs, such as some of the sea snails.

◄ The tentacles, with their suckers, of an argonaut, a relative of the octopus. The female argonaut makes a paper-thin shell. In this she lays her eggs using it as a sort of cradle.

◄ The rounded body of an octopus surrounded by its eight writhing tentacles. Its siphon can be seen at the centre.

► A deep-sea squid. Its body and tentacles are covered with light-producing organs. These appear as dark spots when the lights are switched off. The lights from them are possibly used to attract prey.

ECHINODERMS

Echinoderms are a group of animals found only in the sea. Their name comes from Greek words meaning 'spiny skinned'. In fact not all have spiny skins, but most of them have at least rough skins.

Echinoderms also have an unusual shape. Most animals are bilaterally symmetrical. That is, if you draw an imaginary line lengthwise through its body the two halves of the animal are roughly alike. This means that the head is in front and the tail is behind. There are, however, some animals that are radially symmetrical. That is, the separate parts of the body are arranged in the same way as the spokes of a wheel. So there is no such thing as a head end, or a tail end. The starfish is a good example of radial symmetry.

▼ A brilliantly-coloured six-armed starfish. Starfish can cast their arms and each one will then grow into another starfish.

▼ The strong shells of a bivalve (two-shelled) mollusc are no defence against a hungry starfish. This starfish has gripped each of the shells and is slowly pulling them apart. When they are open, the starfish will push its stomach between the shells and digest the mollusc.

Starfish

The common starfish has five arms. Its upper surface is rough because it has grains of chalk embedded in the skin. On the underside of each arm is a groove. In the middle where these grooves meet, is the mouth. Each groove is lined with tiny suckers, known as tube-feet. When a starfish is moving from one place to another, one of its arms takes the lead. Then, all the tube-feet, on all five arms, act together. They take hold of whatever the starfish is on and, all working together, pull the starfish along.

The starfish feeds on bivalves (page 18). When a bivalve senses a starfish coming towards it, it pulls its shell tightly shut. The starfish wraps its arms round the shell, fastens itself in position with its tube-feet and pulls the shell open. Then it pushes out its stomach, wraps it around the mollusc's body and slowly digests it.

If a starfish eats something that upsets its stomach, it has an easy way to cure this. It simply gets rid of its stomach by casting it out and growing a new one.

Large crabs are one of the chief enemies of starfish. They cut off a starfish's arm with their claws and eat it. But this does not kill the starfish as it can grow a new arm – just as an earthworm can grow a new head or tail end.

Sea-urchins

A close relative of the starfish is the sea-urchin, or sea-hedgehog. The body of a sea-urchin is a round box made of plates of chalk that fit together like a mosaic. The surface of this box is covered with long spines. A sea-urchin looks most unlike a starfish but scientists know the two are related because both have tube-feet.

▶ A crown-of-thorns starfish feeding on coral. These starfish can do enormous damage to reefs, killing the coral and so destroying the homes of many other sea creatures.

▼ A brilliantly coloured and patterned sea urchin from the island of Palau in the Pacific Ocean.

▲ A sea-cucumber climbing over coral rock. The circle of feathery feelers at the front end may sometimes be used to catch small animals swimming by.

Sea-cucumber

Another relative of the starfish is the sea-cucumber. It looks very like the vegetable of that name. It feeds on particles of decaying plants and animals in the mud of the sea-bed. The sea-cucumber also crawls about on its five rows of tube-feet. Like the starfish it can regenerate lost parts. When pursued by an enemy it can squirt its insides at the enemy and grow a new set.

▼ This centipede can run fast. It has to in order to catch the insects on which it feeds.

MILLIPEDES AND CENTIPEDES

There are 8,000 *species* of millipede and nearly 3,000 species of centipede in the world today. The word 'millipede' means 'thousand feet', but only the largest have anything like that number. Most millipedes have a few hundred legs at the most. It is the same with centipedes. Their name means 'hundred legs', and although few centipedes can claim that number, there are some centipedes that have as many as 177 pairs of legs.

◀ Millipedes have more legs than a centipede. Their name means 'thousand legs' but no millipede has as many as that.

▲ Centipedes have powerful claws each side of the mouth. (You can see the one on the left of this centipede's head.) The claws give out a poison that paralyzes the centipede's prey.

The smallest millipedes are only 1.5 mm ($\frac{1}{16}$ in) long. The largest, the African snake millipede, is 28 cm (11 in) long and 2 cm ($\frac{3}{4}$ in) in diameter. Most centipedes are 3–4 cm ($1\frac{1}{4}$–$1\frac{1}{2}$ in) long, but the largest, the giant centipede of Brazil, is nearly as large as the largest millipedes – 26.5 cm ($10\frac{1}{2}$ in) long.

Millipedes and centipedes are often spoken of together. Indeed, they used to be classified together as myriapods, a name which means 'countless legs'. Yet they are not very closely related, although at first glance they look alike. Both are long and slender and have ringed bodies, like earthworms. But their bodies are covered with a tough cuticle, similar to an insect's. They also have jointed legs like insects and a pair of *antennae* on the head.

Not so similar after all

There are, however, several important differences between the two animals although they are both found in soil, especially where there are dead leaves or rotten wood. The first difference is the food they eat. Millipedes feed on decaying plants. Centipedes feed on small animals such as insect larvae and millipedes.

A second difference is that millipedes move only slowly and when disturbed they simply coil up. They depend for protection on the strength of their hard outer skin or *cuticle*. They also have stink glands on their legs. Centipedes, on the other hand, run when they are disturbed.

Another important difference between the two is that millipedes have a pair of legs on each side of each ring of the body and centipedes have only one. The advantage to the millipede in having double the number of legs is that it gives them a better grip when pushing their way through the soil. The legs of both types of animal are jointed. As a result, scientists *classify* them as belonging to the group of animals known as '*arthropods*' – from Greek words meaning jointed legs.

However, unlike either earthworms or insects, female centipedes and millipedes look after their eggs. Many of them build a sort of mud 'hut' in which to lay their eggs. Then they coil themselves around it to give it extra protection. And in some species the females look after the young when they have hatched out.

SHRIMPS AND LOBSTERS

At first glance shrimps and lobsters do not look much like millipedes or centipedes. But all these animals have the thick, hard, outer skin or cuticle, and the jointed legs, typical of the great group of backbone-less animals called arthropods. Scientists think that the remote ancestors of arthropods were much more like worms, that is arthropods have evolved, or changed, through hundreds of millions of years, from thin-cuticled, simple-legged worm-like creatures, to thick-cuticled, jointed-legged creatures. Shrimps, lobsters and their relatives are a sub-group of arthropods called crustaceans. Unlike centipedes and millipedes, the legs of shrimps and lobsters are big in proportion to the rest of the body. Also some of the legs on the front part of the body have evolved into claws.

▼ Shrimps are small crustaceans that live in shallow seas where the sea-bed is sandy. In very deep seas, where no light penetrates, there are shrimps that are dark red. This brightly spotted shrimp lives in the relatively shallow waters of the Mediterranean.

A body in three parts

There is another big change. Instead of a body looking more or less alike throughout its length, as in earthworms, millipedes and centipedes, the body of a shrimp, a crab or a lobster is divided into different parts. There is a head in front bearing antennae, stalked eyes and mouth-parts, including the jaws. Behind the head is the middle part of the body, usually called the *thorax*. Because there is no obvious distinction between the head and the thorax, this front half of the body of a shrimp or a lobster is called the *cephalothorax*. This is from the Greek word *cephalo* meaning 'head'.

The crust covering the cephalothorax forms a continuous cover. On the underside of the thorax are many pairs of walking legs, each of which carries a gill.

The third, hindmost part of the body is the *abdomen*. The rings on this are plain to see. Each ring carries small swimming legs, known as swimmerets. The abdomen ends in a tail shaped like a fan.

► This delicately patterned shrimp has evolved a special relationship with sea anemones, such as the one on which it is resting. The anemones would kill other types of shrimp, but shrimps of this sort are quite safe.

▲ The large brilliantly patterned mantis shrimps are found in shallow, tropical seas. They may be 30 cm (12 in) in length.

▲ The common lobster is found in shallow rocky water along the North Atlantic coast. Lobsters only become red when they have been cooked. Their natural colouring blends well with the rocks and seaweed among which they live.

Sea-dwellers

Shrimps and lobsters are found mainly in the sea, from pools between tide-marks to the ocean depths. Some live in brackish water, and a few even in the fresh water of rivers and lakes. Crayfish, for example, are a kind of fresh water lobster.

Shrimps live mainly on the bottom, lobsters mainly swim just off the bottom. Both swim by flicking the tail, driving themselves backwards. And both eat animal flesh, dead or alive.

'Prawn' is a name given to some of the larger shrimps. It has no scientific meaning because there is no real difference between shrimps and prawns, except for size.

Shrimps may be about 1 cm (½ in) to 30 cm (1 ft) long. Lobsters are larger. The common European lobster is 20–50 cm (8–20 in) long and up to 4.5 kg (10 lb) in weight. The record is 6.7 kg (15 lb). The American lobster may be over 50 cm (20 in) in length and 17–19 kg (38–42 lb) in weight, the record being 20 kg (44 lb).

CRABS

The simplest way to describe a crab would be to say that it is a lobster with its abdomen tucked under its thorax. This is because the thorax which is broader has the small narrow abdomen hidden underneath it. The thorax is encased in a shell that is called a *carapace*.

The 4,000 species of crab that live in various parts of the world today vary widely in size. The smallest are only 6 mm ($\frac{1}{4}$ in) across, while the largest have a carapace 30 cm (1 ft) across. The largest of all is the giant Japanese crab. Although this crab has only a small body, it has very long legs that span 3 m (9 ft).

Crabs live mainly in the sea, from between tide-marks down to the ocean depths, but a few live in fresh water and there are even some that live on land. The robber crab of south-east Asia is well known for climbing palm trees. Some others burrow in mud or sand. They feed mainly on the flesh of other animals, dead or alive, and a few eat plant food such as seaweed. They are, indeed, scavengers, but some crabs, with large claws, will crush living mussels and other such shellfish and make a meal of the soft body inside.

The scavenging nature of crabs can be illustrated by the story of the ship's cook who threw overboard an old saucepan with remains of food still on its inside. His ship was anchored at the time in calm shallow water. In a very short time crabs were coming from all directions, clambering over the saucepan in a mass, fighting and scrambling to reach the food.

Clever camouflage

This story also brings out how numerous crabs are everywhere in the shallow seas.

◄ A male fiddler crab, of the warm seas, has one claw much larger than the other. It waves this backwards and forwards to signal to the female, and also to warn other males to keep away.

Yet they manage to hide themselves so well that it is often difficult to see them. They escape notice by hiding in holes and crevices in rocks or burying themselves in the sand. When they emerge from hiding their colour helps them blend with their surroundings and their rather irregular shape breaks up the outline of their bodies: both important factors for good camouflage.

To make the camouflage even better, some crabs cover their carapace with pieces of seaweed, a piece of sponge or other marine animals. The carapace has tiny hooks on it on which the seaweed and other animals, such as sea-firs, can be hung. So if you ever happen to see a small patch of sea bottom get up and walk away, you'll know. It reminds us of a soldier in battledress with twigs and leaves on his helmet.

A dangerous way of life

The hermit crab has armoured claws and walking legs like other crabs. The rest of its body, however, is soft and unprotected by the hard carapace that other crabs have. This makes it an easy prey for any flesh-eating animal. To avoid this the hermit crab pushes its rear part into the empty shell of a sea snail, such as a winkle or whelk. With its soft body protected in this way it can walk about, dragging its 'house' with it.

▲ Hermit crabs have soft bodies. They tuck themselves into empty sea-snail shells, for protection. This shell has a sea-anemone growing on it.

► A Sally lightfoot crab feeds on a jellyfish that was stranded on the sand when the tide went out.

INSECTS

Insects are by far the most numerous types of arthropod. There are at least a million different kinds of insect, more than all the other types of animal put together. They live everywhere in the world, in vast numbers, except in the seas. There are only half-a-dozen or so species of insects that can live in salt water, although there are plenty that live in fresh water. But all insects wherever they live are cold-blooded.

A basic body design

The first is that their name gives us a clue as to what they look like. The name 'insect' is from the Latin meaning 'notched' or 'cut into'. An insect is notched or cut into three distinct parts, the head, the thorax and the abdomen. The head bears the mouth, eyes and antennae. Behind this is the thorax which bears three pairs of legs and, in most species, one or two pairs of wings. Then comes the abdomen.

▶ The pond skater has long legs to help it glide speedily across water. A small pad of bristles at the end of each leg stops the legs breaking through the surface.

▼ A worker honeybee feeding on nectar from a michaelmas daisy. The three parts of an insect's body – head, thorax and abdomen – show up clearly.

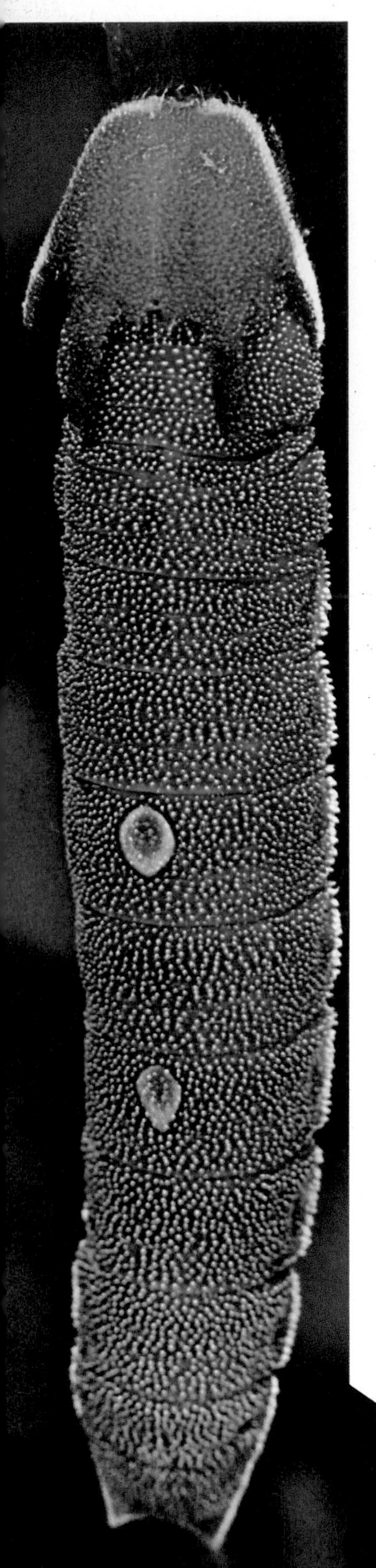

▲ The caterpillar of the two-tailed pasha butterfly will turn into a pupa before becoming an adult butterfly (page 7).

We have seen how the earthworm's body is marked in rings (page 14). The millipedes and centipedes also have ringed bodies and they have many jointed legs. The crustacea (shrimps, lobsters and crabs) also have ringed bodies but they have evolved long jointed legs and claws. Now, in insects, there are further changes still. The most important is that the majority of insects have either one or two pairs of wings.

If we were to lay earthworms, millipedes, centipedes, crustaceans and insects side-by-side, we would have a very good example of what is meant by the word '*evolution*'. This word means 'change'. It would take little change to make an earthworm look like a millipede or a centipede. A few more changes to a millipede or a centipede and it would look like a crustacean. A few more changes and we would have an insect. But evolution takes millions of years.

Variations on a theme

The basic plan of an insect is three parts to the body and three pairs of legs. Add transparent wings to this basic plan and you have a fly. Give it a pair of large coloured wings instead and you have a butterfly. Change the front pair of wings to a pair of hard wing cases covering transparent hind wings and it becomes a beetle. Or else make the abdomen and the transparent wings very long and the insect becomes a dragonfly.

So, no matter how much we change the shape, size or colour, we still have an insect, with three parts to the body and three pairs of jointed legs.

A complicated life-history

Most insects have a complicated life-history and pass through several stages between egg and adult. When it first hatches out a young insect is called a larva. (The larva of a butterfly is called a caterpillar and the larva of a beetle is called a *grub*.) The larvae are quite different from the adults. They have no wings and many species have no legs. They also eat quite different food.

As the larva grows, it gets too large for its outer skin which it has to moult (shed). When full-grown the larva undergoes *metamorphosis*. This word also means 'change' but, unlike evolution, change in a single lifetime. By its metamorphosis, the insect finally becomes an adult. The larva changes its skin for the last time and becomes a *pupa* surrounded by a hard skin or cocoon. This is a resting stage while the insect's body changes to its adult form. Some insects, house flies for example, are pupae for only a few days. Others spend the winter as pupae. But however long this stage lasts, the hard skin finally splits and the adult insect emerges. It takes an hour or so for the body to dry and the wings to expand, but then the insect is ready to fly away.

A few insects, such as grasshoppers and dragonflies have an incomplete metamorphosis – they do not go through all the stages described above. They hatch from the eggs as small wingless versions of the adults, and are called *nymphs*. The nymphs moult at intervals as they grow, and wings appear, getting larger with each moult until the nymph becomes a full-sized adult.

◀ The details of the body and antennae of this bloody-nosed beetle can be seen very clearly in this close-up photograph. The beetle's actual size is 1.8 cm (¾ in).

▶ A brimstone butterfly rests on a cornflower – like all butterflies it has undergone metamorphosis before reaching this adult stage.

SPIDERS

▲ A garden spider has just wrapped up a fly that was caught in her web. She is carrying it to the centre of the web to eat it.

Some people think that spiders are insects, but they are not. Insects and spiders are arthropods, just as lobsters and millipedes are arthropods, and there are many differences between them. (There are, too, only 14,000 species of spider compared with more than a million species of insects.)

Instead of a body divided into three parts, as in an insect, a spider's body is made up of two parts. In front is a small cephalothorax connected by a narrow waist with a large abdomen. The abdomen is much larger in female spiders than in males.

Also, a spider has four pairs of walking legs, instead of the three pairs an insect has. Spiders do not have large compound eyes, like insects. They have only simple eyes. And at the rear end of the abdomen, on the underside, spiders have spinnerets for weaving silk.

Silk for many purposes
It is as spinners that spiders have excelled. Many insects use silk, when they are at the caterpillar stage. But none uses it for as many purposes as spiders.

Not all spiders spin the beautiful geometric webs we see in the garden. Some spin sheet webs, like the cobwebs we see in houses. Others spin funnel webs. The jumping spiders do not spin webs but they do lay down a dragline wherever they go. This is their safety line should they lose their footing. Many other spiders lay down a dragline wherever they go. And all spiders spin a cocoon of silk in which to lay their eggs, to protect them.

All spiders are carnivorous. That is they feed on other animals, especially insects. Web spiders catch their prey by waiting for it to fly into their web. Then they wrap it in silk. Certain tropical spiders hang head upwards at night with a small square of silk held between the hind legs. When an insect flies near it, the spider uses the silken square almost like a butterfly net to capture its prey.

Some water spiders use silk to give them an air supply. They spin webs under water in the shape of a bell. They carry bubbles of air down to the bell where they can use it for breathing, so they do not have to come to the surface so often.

Another way in which silk is used is in courtship. The female spider is larger than the male. When it comes to breeding

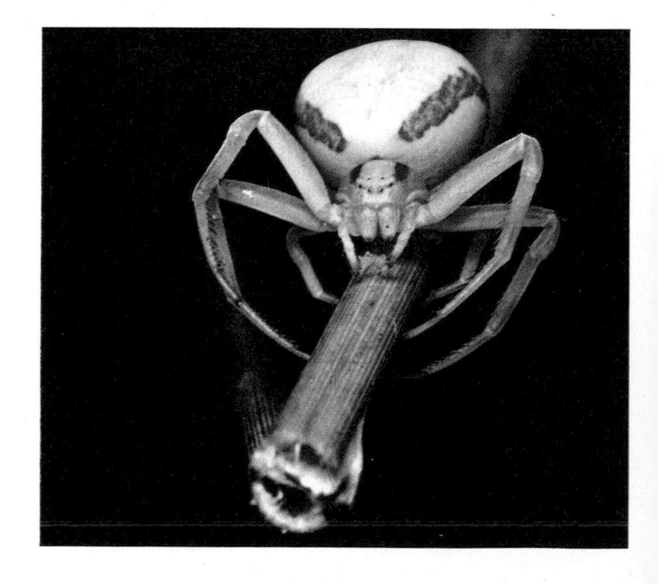

▲ A strikingly patterned male *Micromata roseum* spider photographed at the end of a grass stalk.

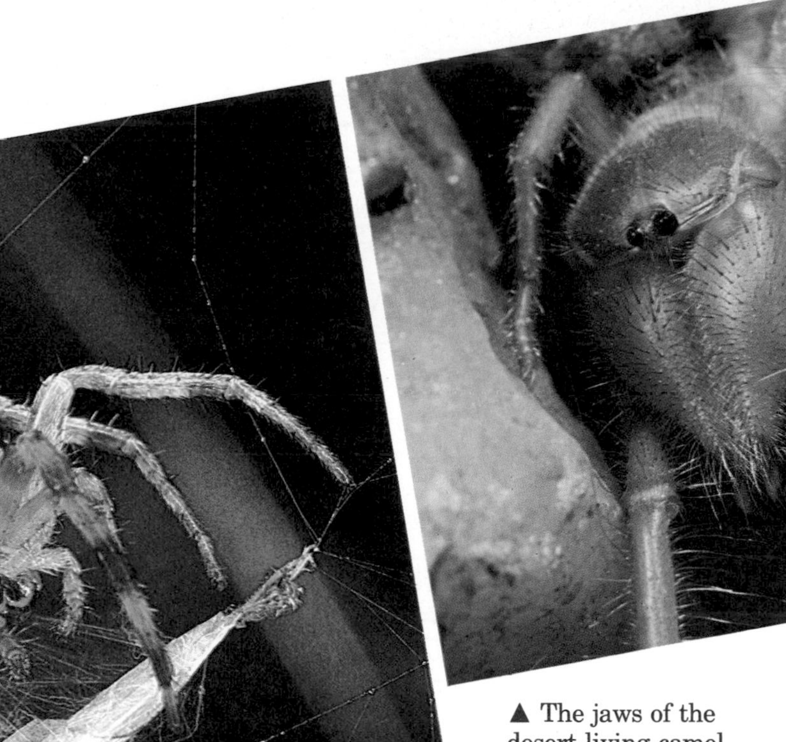

▲ The jaws of the desert-living camel spider look enormous in close-up. This spider is, however, harmless to humans.

▲ This female spider is wrapping her prey in silk so it cannot move. She revolves the victim in her legs, wrapping it in strand after strand of silk.

▲ A mass of newly hatched young spiders, or spiderlings. They will make long silk threads and the wind will carry them to new areas.

time the male spider goes through an elaborate courtship. He does this because the female is apt to eat him once he has fertilized her. Some male spiders use strands of silk to tie the female down. By the time she has freed herself the male will have mated with her and made his escape.

On sunny days in late summer, young spiders can be seen floating through the air on silken threads. Spiders lead solitary lives, coming together only for mating. So, when a batch of eggs hatches the spiderlings must disperse as soon as possible. Each climbs to a high place, lets out a strand of silk and waits for the breeze to lift it into the air. Then, like parachutists, the spiderlings sail through the air until they land on a plant or float down to the grass.

◀ The eight legs that all spiders have are clearly visible in this picture of a baboon spider stalking over the forest floor.

5 SHARKS, RAYS AND BONY FISH

SHARKS

With sharks and their relatives, the skates and rays, we come to the cold-blooded animals with backbones – the *vertebrates*. All the creatures mentioned so far have been *invertebrates* – animals without backbones.

Unlike most kinds of fish, sharks have a skeleton not of bone but of cartilage, or gristle. The bodies of bony fish are covered in scales, whereas the bodies of cartilaginous fish are covered with *denticles*. The word denticle means 'a little tooth'. The denticles of sharks, skates and rays are each made up of a pulp cavity with a layer of dentine around it. This, in turn, is covered with enamel as in the teeth of other vertebrates, including humans.

The teeth in the jaws of cartilaginous fish are also quite different. They are set in rows which gradually move outwards. The worn-out teeth on the outside of the jaw are replaced by new teeth from the inside of the jaw.

Cartilaginous fish have five pairs of gill-slits. These are not covered with a gill-cover, as in bony fish, and are quite easy to see as openings on the body.

There are 250 species of sharks. And although sharks have the reputation of being dangerous to humans, only about 25 of these species have been known to attack humans who are bathing or have accidentally fallen into the sea. The smallest sharks, the midwater sharks, are only 30 cm (1 ft) long. The largest, the great white shark and the whale shark, can be over 13.5 m (45 ft) in length.

The great white shark is the most dangerous to human beings. It feeds chiefly on fish, both alive and dead, and other sea creatures. Most sharks have a similar diet. In contrast the basking shark and the huge whale shark both feed on very small plants and animals, known as *plankton*. They filter the plankton from the sea.

▼ The ferocious teeth and jaws of the great white shark. This shark is the most dangerous of all sharks to human beings, often attacking unwary swimmers in the warm seas where it lives.

▼There is no doubt about why the hammerhead shark got its name. Like most sharks, the hammerhead is a fierce hunter and has been known to eat other hammerhead sharks.

◀ The wobbegong or carpet shark of Australia lies in wait on the seabed for its prey, the patterns on its body blending well with its surroundings.

▼ Some of the shark's relatives are called rays. They are flattened like the skate but the outline of the body is more rounded. One, known as the electric ray, can kill its prey with an electric shock.

RAYS AND SKATES

Rays and skates are relatives of the sharks. They have flattened bodies and spend most of their time on or near the sea-bottom. They feed on clams and similar shellfish. The electric rays are 1–2 m (3–6 ft) long. The muscles of their large fins have special nerve cells made in a similar way to an electric battery. The electric shock these rays can give is extremely powerful.

The largest of the rays is the manta, which lives in tropical seas. It can grow up to 6 m (20 ft) across and weigh 1 tonne (1 ton). These rays occasionally leap out of the sea, a fact that has given rise to the stories of them leaping onto, and sinking, boats.

▶ The giant manta ray provides a free ride for a group of remora fish. Each of the smaller fish has a sucker on the top of its head with which it attaches itself to the ray. Although very large, the manta ray feeds on microscopic plankton.

BONY FISH

▲ The mudskipper lives at the edge of the sea off Malaysia. It spends much time on land, skipping over the mud when the tide is out. For this it uses its muscular tail as well as the front fins that are shaped like legs.

The true or bony fish number at least 20,000 species, in the seas, lakes and rivers everywhere in the world. They all have skeletons of bone. Their bodies are covered with scales. Their gills are covered with an *operculum* or gill cover and their delicate fins are supported by slender fin-rays.

Fish occupy a very wide range of *habitats* and come in a bewildering diversity of shapes and sizes. Think of the shape of a herring or a salmon and compare it with the sea-horse, looking like the knight in a set of chessmen. Or compare the frogfish, with its huge mouth, lying in wait to snap up small fishes as they swim by, to the flounder, the flatfish lying on the sea-bed. There are fish living in underground caves, fish that spend their lives in deep wells, in hot springs or in mountain torrents. There are even flying fish that swim near the surface and leap from time to time into the air to glide like airplanes, as well as deep-sea fish that never see the light of day and carry their own lights in the eternal blackness of the oceans' depths.

Fish are ideal subjects for the study of what is meant by the phrase 'cold-blooded'. There is far less blood present in the body of a fish than in the higher vertebrates, the reptiles, birds and mammals. Moreover, it flows only sluggishly through the fish's arteries and veins. As a result the body of a fish is very little warmer than the water in which the fish is living.

▲ The puffer fish has an excellent defence against its enemies – it blows itself up into a ball so that it cannot be swallowed.

▲ The imperial angel fish lives in reefs in the tropical waters of the Indian and Pacific Oceans. The brilliant colours and strong patterns of many tropical fish provide remarkably effective camouflage against the background of bright corals among which they live.

More exceptions to prove a rule

In other words, fish are cold-blooded animals. Even so there are exceptions. Thus, the fish known as tunny or tuna, and its relatives the albacore and the swordfish, have very muscular bodies. They are very active fish and they are fast swimmers. When muscles work hard they make heat, so that these fishes' muscular bodies are at temperatures noticeably higher than that of the surrounding water.

At the other extreme are the fish that live in very cold waters, such as the Alaska blackfish. For many years it was believed that this little fish, a relative of the pike, could be frozen in a solid block of ice yet emerge unharmed when the ice melted. This is not wholly true but it is the case that the Alaska blackfish and a few others the icefish (page 7) for example, can survive in waters almost at freezing point.

Indeed, all fish, with the possible exception of those living in the hot soda lakes of Lake Magadi in Kenya, can endure cold better than extreme heat.

▲ These sweetlips and blue striped snappers are just two of some 20,000 species of bony fish. As the name suggests, these fish have skeletons of bone. Sharks, on the other hand, have skeletons made of gristle or cartilage.

▼ A flying fish with its 'wings' spread. The wings are, in fact, just the ordinary fins that have developed in a different way to the fins of most other fish.

► It is hard to imagine anything less like a typical fish than the sea-horse. Among its many peculiarities is its tube-like snout, with which it sucks in small swimming animals. Another is its flexible tail, which it can coil round seaweeds to anchor itself.

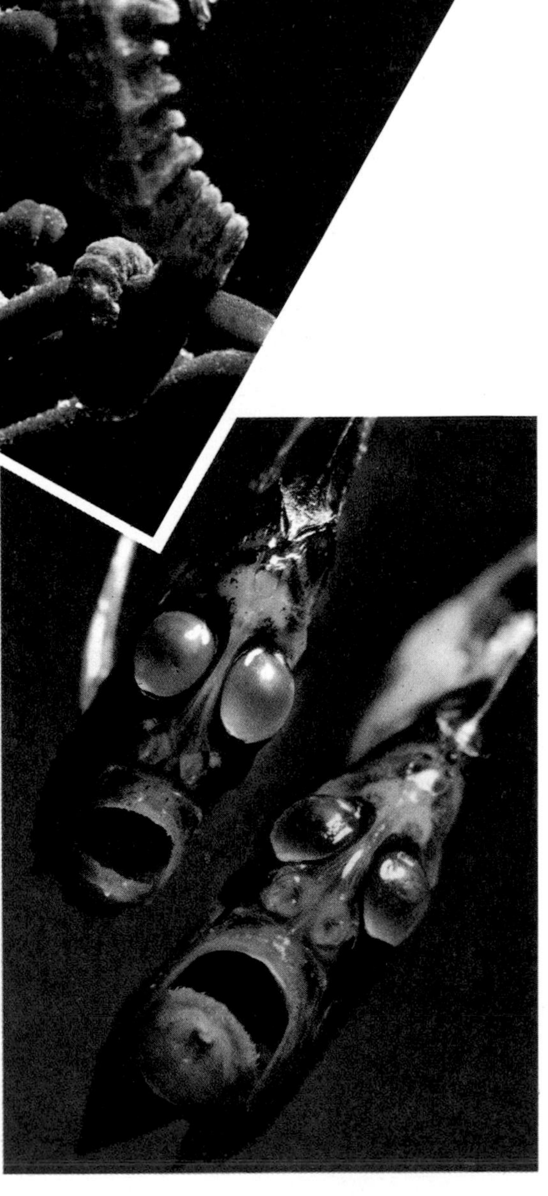

▲ The eel is hatched from an egg laid in the sea. It lives in the sea for its first year of life. Then it swims up a river to a lake, lives there for several years and finally goes back to the sea to breed and die.

▲ Hatchetfish live at great depths. Their huge eyes make the best use of what little light there is and their scoop mouths catch food debris floating down from above.

► By contrast with the eel, salmon are hatched in rivers. They go down to the sea to spend several years feeding and growing. Then they go back to the stream in which they were hatched and go upstream to breed. After that they die.

▼ The dab begins life as a normal fish. Then, early in life, it takes to lying on its side. One of its eyes travels round the head to the side which is to become the upper side of the dab when it has changed into a flatfish.

NEWTS AND SALAMANDERS: AMPHIBIANS WITH TAILS

Amphibians are cold-blooded vertebrates related to fish. Indeed, scientists believe that amphibians are descended from fish. In studying fossils scientists have been able to trace step-by-step how some fish, hundreds of millions of years ago, came out of water and started to live on land. These pioneers were able to do so because they had exchanged breathing with gills for breathing with lungs. They also had exchanged legs for fins. None of these changes took place suddenly – they happened over millions of years.

One of the earliest fossils in this step-by-step sequence of evolution looks like a giant salamander. Its descendants, the 1,500 species of newts and salamanders living today, are cold-blooded animals that readily become sluggish at low temperatures. Those living in temperate regions hibernate during the winter. They have smooth, scaleless skins, four legs and a tail nearly as long as the body.

▼ Newts eggs are laid one at a time on water plants. From them will hatch the larvae which look very like the adults except that they have gills and are much smaller.

▼ The smooth newt is 10 cm (4 in) long, including the tail. This one was photographed during the breeding season. At other times of the year it is a dull olive-brown.

A double life

The word 'amphibian' is from a Geek word meaning 'a double life'. And the animals described as amphibians are equally at home in water or on land. Some tailed amphibians, as salamanders and newts are called, live all their lives in water, coming out on land only for short spells. Others spend all their lives on land, hiding in damp places to prevent their skin drying up. But wherever they live, all amphibians need water in which to breed, and salamanders and their relatives the newts are no exception. The females lay their eggs in water. From these hatch the young, the larvae as they are called. The larvae look like the adults except that they are much smaller. Also they breathe through gills. As they grow they develop lungs. Then they shed their gills and are ready to come out on land. Even then, they still have to keep to moist places and may return to water to keep their skins moist.

An ancient myth

There is an ancient, and false, legend that salamanders can walk through fire and emerge unscathed. Centuries ago this was believed to be literally true. One wonders how many of these unfortunate animals met their deaths as people placed them in hot fires in order to test the idea.

▲ Crested newts live in ponds and streams of Europe. About 13 cm (5 in) long, they spend much of their lives on land, returning to water to breed. It is only then that they grow their crests.

► The European fire salamander, or spotted salamander, must keep its skin cool and moist at all times. Like all newts and salamanders it breathes largely through its skin.

◄ This pale water animal is called an axolotl. It lives in fresh water in Mexico and the western United States. It is really an overgrown larva, for it may grow to 25 cm (10 in) long but still have larval, feathery gills. Unlike most larvae, however it is able to breed.

FROGS AND TOADS

Frogs and toads are close relatives of salamanders and newts, although they look very different from them. They are amphibians too, but they do not have long tails except when they are larvae. They have rounded bodies, large mouths and prominent eyes. They also have much longer legs. The hind legs especially are long and strong and are used for leaping.

Most frogs and toads lay their eggs in water. From the eggs hatch the larvae, known as tadpoles. The name 'tadpole' means round head. When newly hatched a tadpole seems all head and tail.

At first the tadpole feeds on plant food. As it grows larger legs begin to grow out from its body; first the hind legs, then the front legs, so that the 'round head' is, in fact, both head and body.

As the legs grow the tail grows smaller. At the same time the tadpole takes to eating animal food. The gills with which it breathed at first begin to grow smaller and finally disappear. The tadpole now breathes with lungs.

In the end the tadpole has changed to a four-legged, tail-less froglet that leaves the water to live on land.

This change, or metamorphosis, to use the scientific term, from a fish-like tadpole to a four-legged frog, helps us to imagine how a fish could have evolved into a land-living amphibian. This gradual change, which takes place over a few weeks, helps us to understand how in millions of years a fish could have changed into a newt or salamander and later into a frog or a toad.

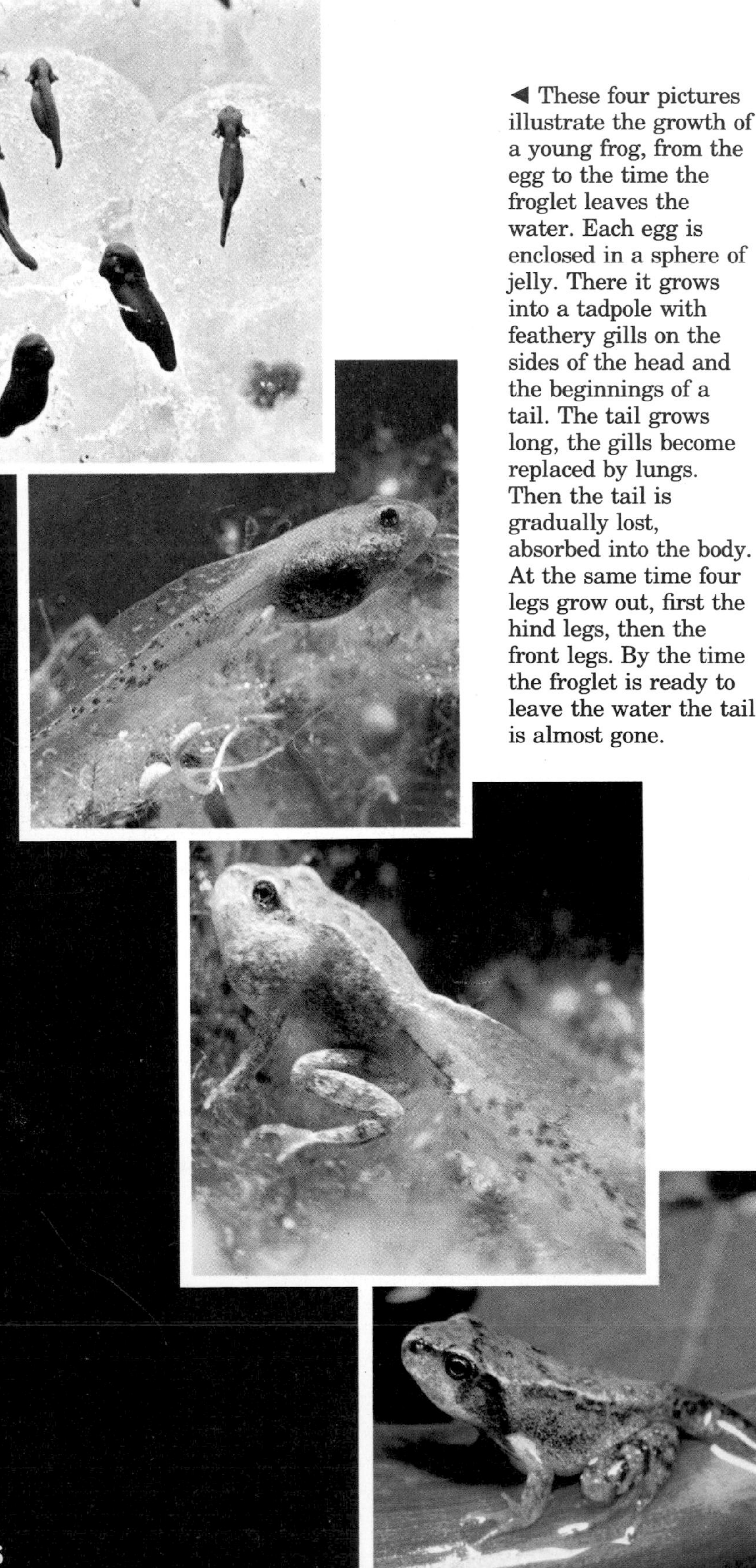

◀ These four pictures illustrate the growth of a young frog, from the egg to the time the froglet leaves the water. Each egg is enclosed in a sphere of jelly. There it grows into a tadpole with feathery gills on the sides of the head and the beginnings of a tail. The tail grows long, the gills become replaced by lungs. Then the tail is gradually lost, absorbed into the body. At the same time four legs grow out, first the hind legs, then the front legs. By the time the froglet is ready to leave the water the tail is almost gone.

Breathing through the skin

There is another way in which amphibians are, even today, halfway between a fish and a truly land-living animal. That is in the way they breathe.

Although amphibians have lungs these are only small, and they breathe mainly through their skin. Some also breathe partly through the skin lining the mouth. There are also lungless salamanders and these breathe entirely through the skin. In contrast there are other tailed amphibians in North America, known as mudpuppies, that keep their gills throughout their life, but have small lungs as well.

Breathing through the skin is possible only so long as the skin is moist. To ensure that it is, there are glands in the skin of amphibians that give out a clear fluid. This spreads over the skin and keeps it moist.

▲ A South American horned frog lies in wait for its prey, to pounce on and seize its victims. It eats insects, small frogs, lizards, snakes and small birds and mammals.

► The midwife toad of south-western Europe has unusual breeding habits. It mates on land and lays its eggs on land. The male carries the eggs on his hind legs. He must make sure the eggs are kept damp, so he goes to water from time to time to wet them.

► The Surinam toad of South America has no tongue, with which other toads and frogs catch their prey. Instead it catches prey with its front feet, pushing the prey into its mouth with its feet.

► The American spadefoot toad lives in the south-western United States. It comes out only at night. During the heat of the day it buries itself in the sand using its spade-shaped hind feet to dig itself backwards into the ground to keep cool.

▼ The bullfrog is a large American frog that calls with a sound not unlike the bellowing of a bull. The female bullfrog lays as many as 20,000 eggs at a time.

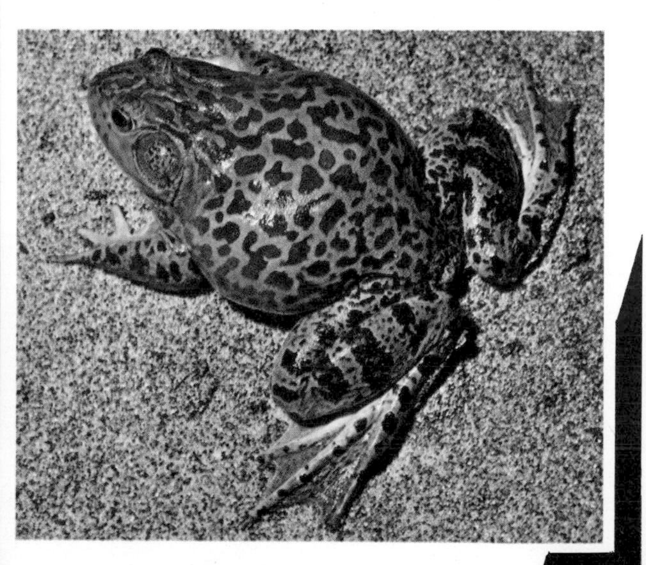

▼ A frog's muscular hind legs enable it to leap well, as this edible frog is doing. Notice, too, how streamlined it looks.

7 REPTILES

TUATARA

The dormouse is famous for its slowness in living and its readiness to fall asleep. In both these things, however, it is pushed to second place by the tuatara, a lizard-like reptile now found only in New Zealand. Even there it is rare. It is fully protected by law, as it should be, since it is one of the most interesting of what are known as living fossils and it is also an endangered species.

The tuatara is the only surviving species of a group of *reptiles* that was numerous and widespread 200 million years ago. It has remained practically unchanged for 180 million years. It is 60 cm (2 ft) long and although it looks like a lizard it is very different in the details of its body. Moreover, most lizards move quickly but the tuatara moves slowly. It grows slowly and does not breed until it is 20 years old and lives to 100 or more. It is believed that some tuataras have lived for 300 years.

It lives in burrows made by seabirds, although it can dig its own. It feeds on insects such as moths, beetles and crickets, which it chews slowly, even at times falling asleep while eating.

A tuatara breathes once every seven seconds, which is even slower than a hibernating lizard. One scientist watched a tuatara and saw no sign of breathing for an hour.

Its mating is also unusual. It pairs in February but the female does not lay eggs until the following October to December. She then scoops out a shallow nest in the ground and lays 5 to 15 eggs. The eggs, however, do not hatch until 13–15 months later.

► The tuatara of New Zealand is famous because it is a living fossil. It looks like a lizard but belongs to an entirely different group of reptiles. The tuatara is even more famous because of its third eye. It was the first animal in which this pineal eye, as it is called, was studied.

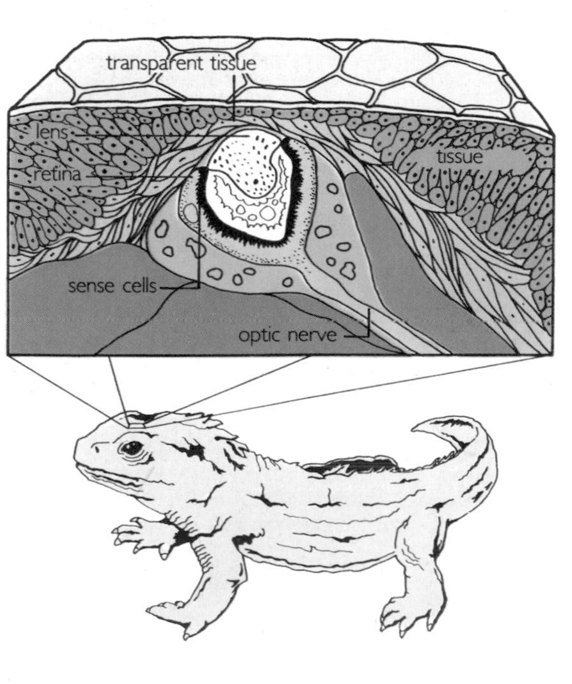

► Diagram of the pineal eye of the tuatara. This 'third eye' is on top of the reptile's brain, just beneath the skull. It has a lens, a retina and an optic nerve. Although the eye is sensitive to light is cannot be moved or focused. It is still a mystery what purpose it serves.

A third eye

This reptile is particularly remarkable for the remains of a third eye in the top of its head, although nobody has been able to say what this is for. This eye is known as the pineal eye. Possibly it acts as a register of solar radiation and controls the amount of time a tuatara can spend in the sunlight. Normally the animal is active only at night but it will sometimes bask in the sun, mainly in late winter or early spring when the sun's heat is not very strong.

Surprisingly tuataras hibernate only lightly and can remain active at temperatures of 7°C (45°F) when other reptiles would become torpid.

CROCODILES AND ALLIGATORS

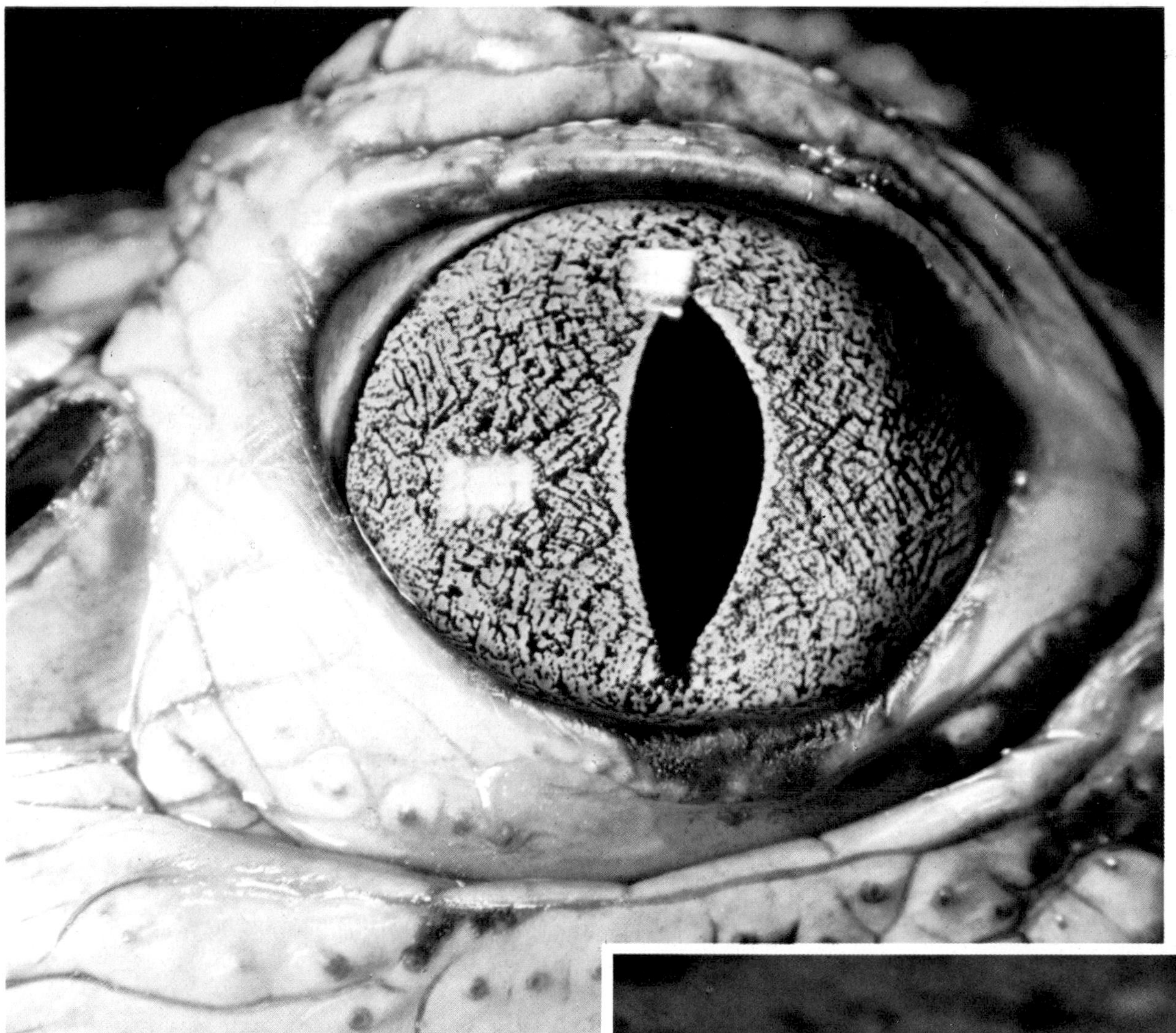

◀ Close-up of a crocodile's eye. The eyes are positioned on the top of the animal's head (**below**) so that when it is submerged the eyes are still above water and it can see its prey, without being seen.

▼ The crocodile's immensely powerful jaws and strong, peg-like teeth make it very easy for the animal to drag even large antelope down into the water and drown them.

Crocodilians look like, and are, relics from the past, survivors of the age of dinosaurs, when reptiles ruled the earth. The largest living today, and the only one to enter the sea, is the saltwater crocodile of Australia. This used to reach 10 m (30 ft), possibly more. A hundred years ago, crocodilians were numerous in rivers in all the warm countries. During the past century they have been shot without mercy for their skins. Large individuals were prime targets in this slaughter. Big crocodiles have therefore been wiped out and today maximum sizes cannot be confirmed.

To anyone but an expert most crocodilians look very much alike. The barrel-shaped body ends in a long powerful tail with long rows of triangular scales which make the thrashing tail an effective and dangerous weapon. Their four legs end in four or five toes, partially webbed. The body is covered with prominent scales. And in front is a large head with a gaping mouth armed with strong peg-like teeth.

The 21 species of crocodilian are divided

▲ Caimans (**above**) are found only in North and South America. By comparison with other crocodilians they are quite small. The young Nile crocodile (**top**) has a lot of growing to do before it reaches its full size. Birds of prey are some of its enemies.

into crocodiles (from the Latin *Crocodylus*), alligators (from the Spanish *el lagarto* meaning a lizard), cayman or caiman (from a Carib word) and gavial or gharial (from the Hindustani).

Crocodiles have the fourth tooth in the lower jaw fitting into a notch in the upper lip. In alligators this fourth tooth is concealed, when the mouth is shut, in a pit in the upper jaw. Caimans are small alligators, about 1 m (4 ft) long, mainly South American, whereas alligators are North American with the exception of the Chinese alligator. The gharial differs from the other three in having a long, slender, beak-like snout. It is a fish-eater, while the rest of the crocodilians eat many different kinds of animal.

Dangerous hunters

What makes crocodiles and alligators particularly dangerous, apart from their size and the strength of their jaws, is that their eyes and nostrils are on the top of the head. This means that they can lie submerged, with only their eyes and nostrils visible above the water. Their victims are, as a consequence, taken unawares until the strong jaws snap shut around them. The reptile can then submerge fully and drown its prey, perhaps an animal coming down to drink, or a person coming to draw water.

On land, a crocodilian normally crawls on its belly with the tail dragging on the ground. It can, however, raise itself on fully stretched legs and run at a fair speed.

Watchful mothers

Reproduction is by eggs, which are an elongated oval, with a hard shell. Depending on the size of the female, 20 to 90 eggs are laid in a pit dug in sand or in a heap of rotting vegetation. Although there is no incubation, the female remains on guard near her eggs until they hatch. A few years ago it was discovered that the newly-hatched crocodilians are carried to water in the mouth of the mother, who may be assisted by her mate.

▲ A large alligator resting in the Everglades National Park, Florida.

► An alligator swimming in the Mississippi river in the USA.

◄ Compare this full-sized Nile crocodile with the head of the young one at the top of the page.

TURTLES AND TORTOISES

Turtles and tortoises are both reptiles that have survived almost unchanged since the Age of Reptiles, which began nearly 300 million years ago and ended 150 million years ago. They live mainly in the tropics, subtropics and the warmer parts of the temperate regions, feeding mainly on vegetation, occasionally smaller animals or carrion (dead animals). There are about 35 species, including giant tortoises a metre or more long and the large leathery turtle or leatherback 1.8 m (6 ft) long. Characteristically, these are slow-moving, heavily armoured animals. The body is enclosed in a box or shell which in most species is rigid. In many species the head, tail and limbs can be withdrawn into the box.

The top of the box, known as the carapace, is formed from overgrown, widened ribs. The lower part, or *plastron*, is also made up of bony plates. Both carapace and plastron are covered with horny plates or shields, known as *scutes*, which provided the tortoiseshell used for ornaments in previous centuries.

This group of animals includes tortoises, turtles and terrapins, three names given different meanings in different parts of the English-speaking world. In Britain 'tortoise' is reserved for land-living species, 'turtle' for marine ones and 'terrapin' for most freshwater members of the group, especially small ones. In North American there is a tendency to call them all 'turtles'.

▲ Hermann's tortoise is also known as the garden tortoise. It lives in Mediteranean countries and like all land tortoises can pull its legs and head into its shell.

▼ The European pond tortoise hunts mostly by night.

▶ The Carolina box turtle of North America can, in the space of moments, shut itself completely in its box-like suit of armour. Even when turned over, all it then presents to an enemy is the hard undersurface or plastron.

Reptiles of great age

The outstanding features of these animals is the slow pace at which they live and the great age they can achieve. The deliberate plodding of the tortoise is well known from the ancient fable of the race between the tortoise and the hare. The great ages to which tortoises are said to live are not always easy to prove, but there are at least three candidates for life-spans of 150–200 or more years: the St Helena tortoise, the Tonga tortoise and Marion's tortoise, all giant tortoises.

The St Helena tortoise was taken to the island of that name before Napoleon began his exile there. It is credited with having reached 300 years, but there are no definite records to substantiate this. The Tonga tortoise was presented to the King of Tonga in 1774 by Captain James Cook. It was accorded the rank of chief and given the name of Tu'imalila. It was no doubt well grown then and was still alive in 1949.

The age of Marion's tortoise is the best authenticated. It was taken to Mauritius in 1766 by Marion de Fresne. It lived in the grounds of the Artillery barracks and was accidentally killed in 1918.

Another good record is of a garden tortoise that lived in the grounds of Lambeth Palace, in London, for 120 years. It died from neglect in 1753.

A giant Galapagos tortoise is credited with having reached 177 years, but as with all tortoises there is no register of births and very few of deaths.

▶ The green turtle is the best-known of the marine turtles. It gets its name from the colour of its fat, which is green due to the seaweeds and other marine plants on which the turtle feeds. The green turtle grows to over 1 m (3 ft) in length.

▼ A young ornate terrapin of South America. Terrapins are small freshwater turtles that live partly on land, partly in the water.

▶ A female leathery turtle scoops a nesting-hole in the sand with her flippers. The 'tears' flowing down her face help counteract the dryness of the air.

LIZARDS

▶ Geckos are small lizards living in tropical countries. They are named for the calls they make: *gekko*. They have suckers or pads on the undersides of their toes by which they can cling to smooth surfaces.

Reptiles include crocodiles and alligators, turtles and tortoises, snakes and lizards. All crocodiles and alligators look alike; turtles and tortoises look alike; all snakes look alike. That is, they look sufficiently alike to leave us in no doubt that we are looking at a crocodile, a tortoise or turtle or a snake. It is not so easy with lizards, because they are a much more varied group of animals.

Taken as a whole, the 2,500 species of lizards are of two kinds. There are those that live on the ground and are flattened from above downwards. Others live in trees or in water and have bodies flattened from side-to-side. Over and above this, lizards differ according to where they live. Their range of *habitats* is wide – from humid tropical forests to deserts, from steppes to mountains. Like all other cold-blooded, air-breathing animals lizards become less numerous towards the cold polar regions and are unable to exist where the subsoil remains permanently frozen.

▼ Monitor lizards live in Africa, southern Asia and Australia. They are usually no more than 1 m (3 ft) long. There is, however, a very large monitor, known as the Komodo dragon, which is up to 3.5 m (11 ft) long.

A tail for defence

A prominent feature of lizards is their tail. This is usually long and whip-like, but it may be short and stubby and used for storing reserve fat. A defensive mechanism found in many long-tailed lizards is the ability to cast the tail and grow a new one. At a point near the base of the tail is a line of weakness. The vertebra at that point is all but severed. The nerves and blood-vessels are narrowed. So, with a strong twitch of the muscles, the whole tail comes away, with little, if any injury, apart from the loss of a few drops of blood. The severed part, meanwhile, writhes and whips around in an agitated manner, diverting the attention of a predator while the animal itself scuttles swiftly and safely out of sight.

The commonest lizards are the skinks. These live in tropical countries. They have an armour of overlapping scales beneath each of which is a bony plate. A normal skink has two pairs of legs with five toes on each. In some skinks the limbs are very small. In others the front pair has disappeared. In a few the hind legs are almost gone, which makes the skink look legless and rather snake-like. The amphisbaenas, the glass snakes and the slow-worms, have lost their legs completely and look like snakes.

Different types of lizard

Other lizards that illustrate the enormous variety of this group of cold-blooded animals are the marine iguanas that feed on seaweed, the Aust alian frilled lizards with an enormous coloured ruff used to display at and deter an enemy, and the Mexican horned toad, a lizard with thorn-like scales that squirts blood from its eyes when cornered. There is a 'flying' lizard in Malaya and there are the strange chameleons that shoot out a long tongue to catch insects and can change colour to blend with their surroundings.

Some lizards have a forked tongue resembling that of a snake and a reputation for being poisonous. There are, however, only two poisonous lizards, the Gila (pronounced heela) monster of Arizona, New Mexico and Utah, and the beaded lizard, of Texas, Arizona and

▲ These two plated lizards of southern Africa look like legendary dragons but are only 30 cm (1 ft) long.

► The slowworm of Europe has lost its legs and looks like a snake. It is harmless to humans and feeds on slugs.

Mexico.

Most lizards are small, but the Komodo dragon, of Indonesia, may be 3 m (10 ft) long and weigh up to 140 kg (300 lb).

Food

Most lizards eat insects which they catch by flicking out their long tongues. There are some exceptions, which is not surprising in such a large group of animals. The horned lizard of the south-west USA eats only ants and the Gila monster eats other lizards. A few species, the iguanas for example, are vegetarian and eat only leaves and fruit. Marine iguanas feed on seaweed.

▲ There is only one species of lizard that goes into the sea of its own accord. That is the marine iguana which lives on the Galapagos Islands. Its food is seaweed.

▼ A chameleon photographed in the act of shooting out its tongue to catch a grasshopper (left). Chameleons can grasp flower stems or small branches with their toes. Moving very slowly, one foot at a time, they stalk insects and catch them with their long tongues, which they shoot out like lightning.

▼ The Gila monster is one of the only two poisonous lizards. It lives in the deserts of the south-western United States. Up to 60 cm (2 ft) long, it feeds on smaller lizards.

SNAKES

► The Egyptian cobra rears its head and spreads its hood before striking. This snake is very poisonous. Long moveable ribs swing out to stretch the loose skin of the neck to form the hood.

There are over 2,000 species of true snakes. Snakes are related to lizards and are, beyond reasonable doubt, descended from lizard-like ancestors. Their bodies are long and cylindrical. Only in the boas and pythons is there any trace of limbs and these are confined to a pair of small bones ending in claws where the hind limbs would be.

So snakes can be described as reptiles with worm-like bodies that creep about on their bellies instead of having limbs for locomotion. There are, however, a number of important changes within the body that mark snakes off from their near-relatives, the lizards.

One of the most noticeable things about a snake is its forked tongue. Except when the reptile is completely at rest its tongue is being constantly extruded and withdrawn, testing its environment. It is withdrawn into a sheath in the floor of the mouth and the tips of its two forks pushed into the twin openings of the Jacobson's organ in the roof of the mouth. This organ has been described as a taste-smell organ.

▼ The red milk snake of North America is harmless. Its unusual colouring makes it look like the deadly coral snake.

Gaping jaws and poison fangs

All snakes are carnivorous. Their food consists of birds, mammals, lizards, other snakes, frogs, toads or fish in the larger species, insects, earthworms and other invertebrates in the smaller species. Snakes do not chew their food but swallow it whole. To do this the mouth needs a wide gape. This is possible because the bones holding the lower jaw to the skull are movable. Added to this, the two halves of the lower jaw, instead of being joined in front are held together by an elastic ligament, and so can be pushed widely apart to enlarge the opening of the mouth.

▲ Sea snakes have nostrils set high up on their heads so that they can breathe without having to raise their heads out of the water.

The shape and arrangement of the teeth assist the act of swallowing. They are recurved, that is they slope backwards, and act as grappling hooks. The two halves of the lower jaw can move independently of each other, acting in a see-saw manner to force prey down the

▲ The Indian python kills its prey by wrapping itself around it and squeezing. The python may be 6 m (19 ft) long.

▲ One of the many rattlesnakes or 'rattlers' that live in North America. At the top of its tail is the rattle, made up of segments of dried skin. The sound of the rattle is a warning that a dangerous rattlesnake is nearby.

▶ Snakes, such as this European grass snake, use their highly sensitive tongues to test their surroundings. The tongues flicker in and out of the snakes' mouths sensing any change.

elastic throat.

When swallowing prey larger than the diameter of its own body, a snake might seem to be risking suffocation. However, to avoid this danger, a snake can push the end of its windpipe out of its mouth and so continue breathing.

Some of the teeth of poisonous snakes are enlarged and form fangs. A poison sac in the angle of the jaws gives out the toxin (poison), a straw-coloured liquid. This travels down a tube and through the fang, which is hollow or grooved, into the wound. The toxin of some snakes acts on the blood of the victims, in other species it affects the nerves of the prey. There are also constricting snakes, the boas and the pythons, that throw loops of their powerful, muscular bodies around the prey, tightening the loops, so crushing the prey and preventing it breathing.

COLD-BLOODED BIRDS?

If you take a bird in your handsit feels warm. You can also feel its heart-beat. Birds are warm-blooded and many escape the cold of winter by migrating. Yet for two thousand years there have been stories of birds hibernating. Nobody believed them.

In December, 1946, in a slot-like canyon in the Chuckawella Mountains in the Colorado Desert, an American professor saw a poor-will, a kind of nightjar, in a hollow in the face of the rock. He could hardly believe his eyes, for the poor-wills had all flown south for the wintcr.

The professor and his students watched the bird for ten minutes. It showed no sign of life, even when they stroked it. The party went away. Two hours later they came back. The bird was still there. The professor picked it up. Still it showed no signs of life. It felt light, as dead birds always do. It was cold. Then it yawned, started to whine, opened its eyes and raised its wings high over its back. In the end it flew away.

▼ The several species of nightjars are known by various names, including nighthawk, poor-will and whip-poor-will. All species look and behave very much alike. The poor-will is now known to be able to hibernate.

A year later the professor came back. There was a nightjar in the same spot. It was probably the one he had seen the year before. This time he took its temperature. The bird was colder than the surrounding air. The professor had also brought a stethoscope. He could detect no heart-beat with it. Nor was there any sin of its breathing.

For several years running the professor returned and kept watch on the poor-will in the Chuckawella Mountains. He was able to prove, by ringing it, that it was the same bird.

We now know that on cold nights some hummingbirds go into a deep sleep. They behave like the poor-will. Their body goes cold. Their heart seems to stop beating. There is no sign of breathing. One hummingbird remained like this for two nights and a day.

Sleepy swifts

Swifts in North America have been found to be torpid in winter. This is especially true of the chimney swifts. When taken into a warm room they became active and later flew away.

Then, some years after the poor-will was found asleep in winter, it was noticed that the common swifts of Europe will fly away from a thunderstorm. They fly away at right angles to the path of the storm.

One year, a particularly heavy storm occurred over Finland. All the swifts flew south, several hundred kilometres, and stayed away several days. This was at nesting time and a scientist wondered what could have happened to their nestlings, with no parents to feed them. He found they merely went cold, almost stopped breathing and showed all the signs of hibernation until the parents came back and were able to feed them once more.

So, perhaps the stories of hibernating birds were based on observation and were not just wild fantasies.

▲ Hummingbirds are very small. Small animals lose heat more quickly than large ones. If food is scarce or the temperature drops lower than usual, some hummingbirds temporarily give up their temperature control and become cold-blooded.

◀ If the parent swifts are away for a long time the young swifts risk starvation. So they go into a cold-blooded state until the parents return to feed them.

MAMMALS

▼ A hibernating dormouse rolls itself into a ball and goes so fast asleep it can be rolled along a table without waking. Its body then is cold, just as if the dormouse were dead.

Mammals are warm-blooded animals. That is, the temperature of their bodies remains approximately the same whatever the temperature of the surrounding air. This means they are able

▲ A group of horseshoe bats asleep in a cave. A horseshoe bat when fast asleep hangs by its feet and wraps its wings around its body, like a cloak. Its temperature drops to that of the surrounding air.

to control the temperature of their bodies to keep it constant. There are, however, a few mammals that are able to give up this temperature control at times. They do so when they go to sleep for the winter, or hibernate.

A few examples of mammals that hibernate are the dormouse and the hedgehog of Europe and the woodchuck of North America. Each year in the autumn, or fall, they roll themselves into a ball in a warm nest. Their breathing drops almost to nothing, as does their heart-beat. They seem almost dead. And so they remain for up to six months of the year. In other words, for part of the time they are warm-blooded and for the other part, throughout the winter, they are cold-blooded.

Cold-blooded bats

There are, however, many mammals that go even further and are cold-blooded for at least three-quarters of their lives. These are the bats, that make up about one-sixth of all mammals. This is especially true of the insect-eating bats of the temperate regions of the world.

An insect-eating bat hunts insects mainly at night. Bats can sometimes be seen hunting in broad daylight but that is unusual. As a rule, they come out at dusk to hunt. Even then, they generally hunt for only part of the night, perhaps for an hour after nightfall and again for an hour or so before dawn. For the remainder of the night they rest.

By day, when in their roosts, they no longer have temperature control over their body and become, in effect, cold-blooded. Both their rate of breathing and their heart-beat slow down. Their bodies become cold and they appear lifeless.

This is what happens during the period of late spring, the summer and during early autumn. So for half the year an insect-eating bat of the temperate regions spends more than half its time behaving as if it were cold-blooded. In addition, for the other half of the year, which includes most of the autumn and throughout the whole of winter they 'sleep like the dead'.

Perhaps that is why insect-eating bats live for up to 30 years, which is a long life for so small an animal!

GLOSSARY

Abdomen The part of the body in which an animal's stomach and intestines are found. In insects it is the last of the three parts of the body.
Amphibian Animal, such as a newt or toad, that can live in water or on land, although they need water in which to breed.
Antenna (*plural* **antennae**) Sometimes called a feeler, antennae are used mainly as sense organs for smelling and touching.
Arthropod Short for Arthropoda, meaning 'jointed legs', the scientific name for insects, spiders, crustaceans, millipedes and centipedes.
Carapace The hard upper shell of crustaceans, turtles, tortoises, and terrapins.

▼ A female leathery turtle or leatherback. The leathery turtle may reach nearly 3 m (over 8 ft) long and weigh nearly 1 tonne (1 ton).

Cephalothorax The term used to describe the front part of the body of a shrimp or a lobster, where there is no obvious distinction between the head and the thorax.
Cilia Short hair-like structures projecting from many kinds of living cell, including many single-celled animals.
Classify To arrange in groups, each member of the group being related to the other members. Scientists classify (arrange) animals into groups to show the relationships between them.
Crocodilians Members of the group of reptiles that include crocodiles, alligators, caymans (caimans) and gavials (gharials).
Crustacean Animal such as a crab or a lobster with a hard, crust-like shell (*see* Arthropod).
Cuticle Tough outer covering of, for example, a centipede's body.
Denticle A small tooth-like projection. The surface of a shark's body is covered with denticles.
Evolution The gradual development and change of different forms of life over many generations and thousands of millions of years.
Flagellum (*plural* **flagella**) Long hair-like organs on the bodies of many single-celled water creatures; their lashing motion moves the animals through the water.
Gill Organ with which fish and other water-living animals breathe.
Grub The larva of an insect, especially a beetle, that lives in earth or rotten wood.
Habitat The natural home of a plant or an animal.
Hibernate/hibernation Some animals survive the winter by going into a type of sleep. They become active again when the weather becomes warmer.
Invertebrate Animal with no backbone.
Larva (*plural* **larvae**) An insect from the time it leaves the egg until it becomes a pupa. (If the insect does not change into a pupa it is called a nymph before it becomes an adult.)
Mantle Part of the upper surface of a slug's or snail's body. The mantle produces the lime from which the outer shell of the snail and the internal shell of

◀ The edible sea urchin lives among rocks and seaweeds along the coasts of the English Channel and western Europe. It prefers shallow water and is seldom found further out to sea than about 50 m (160 ft) from low water mark.

a slug is made. The mantle is also used for breathing.
Metamorphosis A stage in a creature's development, when its body changes greatly in outer shape and appearance and also inwardly.
Mollusc Soft-bodied animal belonging to the group that includes bivalves, such as oysters and mussels, slugs and snails, octopus and cuttlefish.
Nymph The young of an insect that does not become a pupa during its development from egg to adult.
Operculum The cover of a fish's gills.
Parasite A plant or animal that takes food and nourishment directly from another plant or animal.
Plankton Minute plants and animals that live in large numbers in water.
Plastron The hard lower part of the shell of turtles, and tortoises and terrapins.
Polyp Individual member of a large colony of nettle animals, such as coral.
Protoplasm Colourless substance that makes up most of the cells of living plants and animals.
Pupa (*plural* **pupae**) The third stage in the life of many insects that comes between the larva and the adult.
Radula The strong file-like organ of slugs and snails.
Reptile Cold-blooded animal such as a crocodile, snake or lizard.
Scutes The horny plates or shields covering the carapace and plastron of turtles, tortoises and terrapins.
Setae The bristles which invertebrates such as earthworms use to get a grip on the ground as they move.
Species A group of animals or plants that look alike and behave in the same way, and can inter-breed.
Thorax The middle of the three parts of an insect's body.
Torpid The sluggish state of an animal that is not hibernating.
Vertebrate An animal with a backbone.

▼ Thread snakes live in tropical America. They are unusual because they have teeth only in the lower jaw.

INDEX

Acknowledgments

Heather Angel, Ardea, Ian Beames, C Bevilacqua/S Prato/M Bavestrelli, Ron Boardman, A Borgioli and G Cappelli, H Chaumeton, Bruce Coleman, G Costa, Gene Cox, PM David, D Faulkner, Giacomelli/Margiocco, D Guissani, Eric Hosking, Archivio IGDA, Jacana, La Palude, R Maltini–P Solaini, A Margiocco, G Mazza, R Merlo, Pat Morris, WJC Murray, NHPA, Nature Photographers, Oxford Scientific Films, Petretti, Premaphotos Wildlife, A Rizzi, John Robinson, Seaphot, TH Tashjian, J Vasserot, MI Walker, John Norris Wood.